T0250509

*Health Promotion
and Interactive Technology:
Theoretical Applications
and Future Directions*

LEA'S COMMUNICATION SERIES
Jennings Bryant/Dolf Zillmann, General Editors

Selected Titles in Applied Communication (Teresa L. Thompson, Advisory Editor) include:

For a complete list of other titles in LEA's Communication Series, please contact Lawrence Erlbaum Associates, Publishers.

Health Promotion and Interactive Technology: Theoretical Applications and Future Directions

Edited by

Richard L. Street, Jr.
Texas A & M University
William R. Gold
Blue Cross Blue Shield Blue Plus of Minnesota
Timothy Manning
Texas A & M University

Routledge
Taylor & Francis Group

LONDON AND NEW YORK

First published 1997 by Lawrence Erlbaum Associates, Inc.

Published 2019 by Routledge
2 Park Square, Milton Park, Abingdon, Oxon OX14 4RN
52 Vanderbilt Avenue, New York, NY 10017

Routledge is an imprint of the Taylor & Francis Group, an informa business

Copyright © 1997 Taylor & Francis

Library of Congress Cataloging-in-Publication Data

Health promotion and interactive technology : theoretical
applications and future directions / edited by Richard L.
Street, Jr., William R. Gold, Timothy Manning.
        p.   cm.
    Includes bibliographical references and index.
    ISBN 0-8058-2204-6 (cloth : alk. paper). — 0-8058-
2205-4 (pbk. : alk. paper)
    1. Health promotion—Interactive multimedia.   2. Pa-
tient education—Interactive multimedia.   I. Street, R. L
(Richard L.)   II. Gold, William R.   III. Manning, Timothy (Timothy R.)
RA427.8.H4926   1997
613'.0285'67—dc21   96-37803
                                        CIP

ISBN 978-0-8058-2204-5 (hbk)

To our families, who contribute the most to our health, well-being, and quality of life

Contents

Part III
The Future of Interactive Technology for the Promotion of Health

Preface

During our childhood years, many of us as budding chemists or cooks experimented with mixing various ingredients together to create something new and wonderful. Our choices for ingredients were not purely random; we often picked things that were tasty, had funny colors, or made bubbles. Although the process of creation was fun and perhaps educational, the resulting concoction more often than not fell short of being a scrumptious dessert or the newest rocket fuel. We had neither a well-designed plan nor an adequate understanding of how the parts we mixed together interacted with and affected one another.

In many ways the use of interactive computing technology for health promotion is much like these childhood escapades. In the mix are several exciting but volatile ingredients that we do not fully understand—the shifting emphasis in health care from treatment to prevention, exponential advances in computing technology, the corporatization of medical and health services, and a public that is increasingly wanting, even demanding, a more active role in health and medical decision-making. We understandably are enthusiastic about what the new communication and information technologies potentially can do for health promotion initiatives. The technology enhances our ability to expand the scope and reach of a health campaign, create dynamic multimedia patient education programs that are tailored to the user's needs and interests, provide access to health information and services, and construct networks of support for people with similar health needs and concerns. Yet, there is also a gauntlet of obstacles. Health care organizations often do not invest in health promotion and educational initiatives, much less in newer technology needed for these activities. Many people cannot afford, have no interest in, or are intimidated by computers. Some of the computer programs and services "on the market" are difficult to use or are not particularly interesting, informative, or effective.

The purpose of this book is to examine how interactive technology is and can be used for health promotion and patient education. Our intent is to present a work that serves as a guiding heuristic for identifying audiences, contexts, and health issues for which programs and services using interactive media will lead to improved health outcomes, particularly when compared to more traditional media such as brochures, videotapes, telephones, public service announcements, and professional consultations. The fundamental premise underlying this book is that the production and implementation of these programs must be based on proven conceptual models that provide the best insight into how to maximize utilization, how to optimize the user's experience within the mediated environment, and how to develop and deliver these programs so that they help achieve the health behavior or outcome of interest. Without a sound conceptual framework, the best of intentions will not reach fruition.

Throughout the book, there are three recurrent themes. First, no longer does good health simply mean the absence of malfunction or pathology of the body. The new orientation toward health is broader and more proactive with an increasing emphasis on disease prevention, on psychosocial aspects of well-being (e.g., quality of life, family health), and on an informed, motivated citizenry that makes healthy and gratifying choices in everyday activities (e.g., diet, using safety belts), in managing chronic disease, and in utilizing health care services. For our purposes, health promotion is broadly construed to refer to programs, services, and interventions aimed at providing people with resources (e.g., information, social support, decision support, expert advice) to satisfy their health-related needs and to help them make appropriate health-related decisions.

Second, interactive technology is perhaps the most promising medium for achieving health promotion initiatives. The technology allows health messages to be individualized to the particular needs or interests of the user, provides access to practically unlimited information (e.g., through the Internet and the World Wide Web), enables contact with other people (e.g., experts, others with similar health concerns) throughout the country and the world, can be used to create games and simulations to foster disease management and prevention skills, and can accomplish these functions using an array of images, sounds, and text.

Finally, changes in health care markets offer new opportunities for using information and communication technology. As health becomes more centered on wellness and prevention and less on disease and treatment, the environment is ripe to demonstrate innovations in health promotion that may lower the costs of care, increase access to health information and services, and help individuals make choices that ultimately will contribute to their health and well-being.

OVERVIEW OF THE BOOK

This book is divided into three parts. "Theoretical Perspectives" presents four chapters that offer conceptual frameworks for examining how characteristics of media, messages, and users relate to one another and how interactive media can effectively and appropriately contribute to improved health outcomes. In chapter 1, Street and Rimal describe various features of interactive technology (e.g., interactivity, multimodality) that are particularly well suited for health promotion and patient education. After reviewing studies indicating that computer-based systems are at times better and at times no more effective than more traditional methods (e.g., brochures, videotapes, and professional consultations), the authors offer a theoretical framework that describes three stages of influence that will determine the effectiveness of health promotion using interactive technology: (a) processes related to the adoption and utilization of the technology, (b) the user–media–message interaction, and (c) other personal, social, cultural, and economic factors affecting health behaviors and outcomes. The chapter by Rimal and Flora (chapter 2) examines how the interplay among a user's personal characteristics (e.g., knowledge, motivation), message content, and media can affect the manner in which an individual processes health messages. The authors focus specifically on the characteristics of interactive media that are conducive to learning, skill acquisition, and the retention of information. By understanding the dynamics of the user–message–media interaction, researchers and program developers will be better able to adapt the technology to achieve health goals.

Skinner and Kreuter (chapter 3) outline several theoretical frameworks that are useful for identifying key variables and processes affecting health beliefs, attitudes and behaviors. Using the PRECEDE/PROCEED model of health promotion planning, the authors integrate various features of these theories to design a heuristic for developing and evaluating health promotion interventions. The authors conclude with an examination of how interactive technology can be used to increase the likelihood of a successful campaign or intervention. Finally, Manning (chapter 4) identifies four roles the health promoter may assume in a campaign—messenger, educator, marketer, and facilitator. He argues that health promotion interventions should accommodate differences among individuals' learning styles, life circumstances, preferences, and needs. By viewing computing as a medium (rather than a tool), the health promoter can create an "environment" that encourages active exploration of the information on a particular health topic. A user who has freedom to move within the environment can develop a sense of responsibility and control as he or she integrates new information with existing health knowledge, needs, and behaviors.

The second part of the book, "Using Interactive Technology to Improve Health" has six chapters from research teams that have developed, implemented, and evaluated specific computer applications for health promotion. Each chapter describes the program and its targeted audience, presents the theoretical rationale for the program, offers evidence of its effectiveness, and concludes with suggestions for the development of similar types of programs. In chapter 5, Hawkins and his colleagues describe their Comprehensive Health Enhancement Support System (CHESS) programs, which are in-home computer systems that enable users to access information services and communicate with experts and others experiencing similar health problems. CHESS has been successfully tested with HIV and breast cancer patients, and piloted for people facing several other health crises. Research to date indicates that people have different styles with respect to how they use the CHESS system, specifically with respect to what services or programs they find most useful and how often they use the programs. Furthermore, CHESS users have reported both better quality of life and lower health care costs than have control groups. The authors offer specific suggestions for increasing the utilization and effectiveness of computer programs and networks designed to provide health information and services.

Lieberman (chapter 6) examines the use of computer "games" to help children and adolescents learn about managing chronic conditions such as asthma and diabetes. Lieberman describes how theories from the fields of psychology, education, and communication have been integrated into the design of a series of health education video games. Using role-playing formats, the games have enhanced players' knowledge about the health condition, interest in good health, sense of self-efficacy, and self-care skills.

In addition, the games provided opportunities for social support among the players and were associated with more frequent communication with parents about health issues. The chapter concludes with a discussion of how the effectiveness and reach of these health games will be furthered improved once they become available on-line through commercial, community, or medical networks.

In chapter 7, Street and Manning discuss the development and evaluation of two interactive multimedia programs for breast cancer education, Options for Breast Cancer Treatment (for women newly diagnosed with early breast cancer) and Options for Better Breast Care (for women who do not routinely screen for breast cancer). We report findings indicating that a carefully designed educational intervention—whether it be a brochure, videotape, or multimedia program—can increase one's knowledge about and interest in the targeted health issue. However, the results also suggest that people who are highly involved in the topic, such as women with breast cancer or who are concerned about the disease, are more engaged by and responsive to interactive multimedia programs (e.g., spend more time with the program, explore more topics) than are people with little interest in the subject. We conclude the chapter with a discussion of how the most effective medium for a health promotion or patient education intervention will depend on available resources, audience involvement in the issue, and specific educational objectives (e.g., increasing knowledge, emotional support, and problem-solving skill).

In chapter 8, Jimison discusses how interactive technology is much more capable of tailoring health messages to an individual's needs, interests, and preferences than are other media. She discusses her research in the development and evaluation of the Angina Communication Tool (ACT), a program for educating patients about treatment alternatives for angina. ACT individualizes the patient education depending on the patient's input (e.g., treatment preferences, valued outcomes) and other medical factors (e.g., seriousness of the disease, age). Programs like the Angina Communication Tool (as well as others discussed in this section of the book) are capable not only of helping users understand their options, they also can contribute to increasing patient involvement in the decision-making process.

The next two chapters (Brennan and Fink, chapter 9, Scheerhorn, chapter 10) examine the use of interactive technology to create computer networks for individuals facing specific health problems. Brennan and Fink's chapter discusses how the range of communication possibilities (broadcast, one-to-one, one-to-many) available on computer networks facilitate social support and create an environment likely to enhance the user's sense of "connectedness." In discussing their work with HealthLink, a network for caregivers of elderly family members, the authors demonstrate how these networks have contributed to the caregiver's emotional well-being and confidence. Similarly, Scheerhorn describes his work with HIGHnet, a computer network for people with hemophilia. Scheerhorn's analysis focuses more directly on issues related to implementation and use of the system. Although still in progress, the research to date has produced two notable findings. First, most users found the program to be of great benefit and used it regularly. Second, certain services were more widely used (e-mail and bulletin boards) than others (informational libraries). Scheerhorn speculates on the implications of such findings when developing health-related networks.

The third section of the book, "The Future of Interactive Technology for the Promotion of Health," examines institutional and developmental issues regarding the use of interactive media for health promotion and the delivery of services, new ways to think about developing these systems, and the role of interactive technology in health care of the future. In chapter 11, Kahn analyzes the dramatic changes that are occurring both in health care and in information technology. He discusses what shifts may occur in the health care "balance of power" among health care professionals, consumers, employers, communities, and citizens. A growing acceptance of a holistic view of health, changes in the delivery of care, and the increasing availability and power of information technology will allow more freedom and provide more resources for people to make their own health care decisions and better utilize health care services. He concludes the chapter with a discussion of Health Action Performance Information Support Systems (HAPISS), an example of how interactive technology may be used in the health care of the future.

Gold's chapter (chapter 12) examines the shift to managed care within the health care industry and how these changes present both barriers (e.g., short-term thinking, institutional inertia) and opportunities (quality assurance, increased emphasis on preventative care) for introducing interactive technology into these organizations, particularly for the purposes of patient education and health promotion. The chapter offers several strategies for dealing with these issues that may help clinicians, administrators, researchers, and program developers more effectively increase institutional adoption and utilization of these programs. Finally, we conclude this volume (chapter 13) with a panel discussion among Jack Wennberg, Dave Gustafson, and Tony Gorry, three of the pioneers in the use of computers to provide health services and patient education. Each of these individuals offers his perspective on the prospects and pitfalls of the technology and provides specific suggestions that should help researchers, clinicians, and program planners in the development and application of their systems and programs.

ACKNOWLEDGMENTS

We would like to express our thanks to several individuals and institutions who helped make this book possible. To Jennings Bryant and Teri Thompson, we are grateful for their encouragement and invaluable support for this project. We also thank Amy Olener, who worked diligently to keep production of this volume on track. The research reported in chapter 7 of this volume received support from the Texas Advanced Research Program (Grant #010366-056). Finally, for their assistance in underwriting conference calls, recording equipment, and duplication costs, we are indebted to the Program in Leadership and Health Care Policy, Center for Public Leadership Studies, Bush School of Government and Public Affairs at Texas A & M University.

ABOUT THE EDITORS
AND CONTRIBUTORS

Richard L. Street, Jr., PhD (University of Texas, 1980), is Professor of Speech Communication and of Internal Medicine at Texas A&M University. He is also Director of the Program in Leadership and Health Care Policy, Center for Public Leadership Studies which is part of the Bush School of Government and Public Service. His research examines communication patterns between patients and health care providers and the ways interactive technology can be used to improve health outcomes. His research has been published in various journals including *Cancer, Family Medicine, Medical Care, Social Science and Medicine,* and *Diabetes Care.* In addition to this volume, he is also coeditor of *Sequence and Pattern in Communicative Behavior* (with Joseph Cappella).

Timothy R. Manning, MS, is Director of Information Technology at Texas A&M University Health Science Center. He also directs the projects of the Information Environments Lab which provides design and technology support for interdisciplinary research projects involving computing and communications technologies. His research focuses on models of interaction in mediated environments and on user experience.

William R. Gold, MD, is Medical Director for Blue Cross Blue Shield Blue Plus of Minnesota. He oversees an initiative called Clinical Systems Improvement. He works with physician groups throughout the state as they transition their organizations to managed care. The activity includes educational programs, clinical process improvement, disease management, and consulting on a variety of related issues. Formerly, he was Director of the Institute for Health Care Evaluation at Texas A & M Health Sciences Center and worked as an obstetrician and gynecologist at Scott and White Clinic in Temple, TX.

ABOUT THE CONTRIBUTORS

Eric Boberg, PhD, (University of Wisconsin–Madison, 1986), is Assistant Researcher in the Center for Health Systems Research and Analysis. His research interests include empowerment of people living with HIV.

Earl Bricker, BA (Western Illinois University, 1975), is Assistant Researcher in the Center for Health Systems Research and Analysis. His research interests include the role of leadership in organizational change governed by quality management and continuous improvement principles.

Patricia Flatley Brennan, RN, PhD, is Moehlman Bascom Professor, School of Nursing and College of Engineering, University of Wisconsin–Madison. Dr. Brennan recently was professor of nursing and systems engineering and sociology, Case Western Reserve University, Cleveland, Oh. Dr. Brennan received a Masters of Science in Nursing from the University of Pennsylvania and a PhD in Industrial Engineering from the University of Wisconsin–Madison. Following seven years of clinical practice in critical care nursing and psychiatric nursing, Dr. Brennan has held several academic positions. She developed and directed ComputerLink, a electronic network designed to reduce isolation and improve self care among home care patients. Dr. Brennan is a Fellow of the American Academy of Nursing and a Fellow of the American College of Medical Informatics. She is Associate Editor for the *Journal of the American Medical Informatics Association.*

Sue V. Fink, RN, PhD, received her PhD in Nursing from the University of Michigan. Dr. Fink's research examines health promotion strategies, and she has investigated the use of health promotion strategies among residents of longterm care facilities. She is Research Professor, University of Michigan School of Nursing.

June A. Flora, PhD, is Associate Director and Senior Research Scientist at the Stanford Center for Research in Disease Prevention at Stanford University's School of Medicine. She has published extensively in the area of health communication, particularly on mass media campaign effects, message design, and audience analysis. Dr. Flora's current research focuses on the development of theory and methods to explain social change in whole communities. She is also conducting experimental research on the influences of advertising on adolescents' normative perceptions and behavior. Dr. Flora's research has focused on a variety of health issues including cancer, cardiovascular disease, nutrition, physical activity, smoking cessation, AIDS/HIV, and most recently, violence among youth.

Anthony Gorry, PhD, is Vice President of Information Technology, Director of the Center for Technology in Teaching and Learning and Professor of Computer Science at Rice University and Adjunct Professor of Neuroscience at Baylor College of Medicine. Dr. Gorry's research concerns the impact of information technology on organizations and society. He led the development of The Virtual Notebook System, a distributed multimedia hypertext system to support distributed teams. He previously conducted research on the application of artificial intelligence in medicine and on the development of decision support systems. Dr. Gorry directs Rice's Center for Technology in Teaching and Learning. The Center is developing computing and telecommunications for sharing knowledge in schools, universities, the work place and the home. Dr. Gorry is also a Director of The W. M. Keck Center for Computational Biology, a joint endeavor of Rice, Baylor College of Medicine and the University of Houston. He directs a training grant on computational biology funded by the National Library of Medicine.

David H. Gustafson, PhD, is a professor of Industrial Engineering and Preventive Medicine at the University of Wisconsin in Madison. He is past chair of the Industrial Engineering Department and founder of the Center for Health Systems Research and Analysis (CHSRA), a multidisciplinary research center employing systems analysis, decision science and decision support technologies to address patient care and health policy problems. Dr. Gustafson holds bachelors, masters, and doctoral degrees in Industrial Engineering from the University of Michigan where he was a W.K. Kellogg Fellow in Health Administration. Current research efforts focus on development and evaluation of CHESS, the Comprehensive Health Enhancement Support System, which uses expert systems, computer mediated communication, data bases and multimedia presentations in a computer-based system to help people facing crises of health, physical abuse, professional failure, and substance abuse.

Robert P. Hawkins, PhD (Stanford University, 1975), is Professor in the School of Journalism and Mass Communication at the University of Wisconsin–Madison. His research interests include attention and viewing styles, and applications of new technologies to health communication.

Holly B. Jimison, PhD, is Assistant Professor of Medical Informatics and Assistant Professor of Public Health and Preventive Medicine at Oregon Health Sciences University. She also serves as Director of the Informed Patient Decisions Group, conducting research involving the development of computer tools to enable patients to be active and informed participants in their medical care decisions. Dr. Jimison received her Doctorate in Medical Information Sciences from Stanford University, with dissertation work on using computer decision user models for the presentation of consumer health information, risk communication, the modeling assessment of patient preferences for health outcomes, and computer-generated explanation of medical decision models.

Gary S. Kahn, MD, MEd, is President of Healthbridge Systems in Encinitas, California and assistant clinical professor, Dept. of Preventive Medicine and Biometrics, University of Colorado School of Medicine. Dr. Kahn has been involved in the use of computers in medicine for over 25 years—especially for educational applications. He is board certified in Family Medicine and Preventive Medicine and was formerly in private practice in Boulder, Colo. He founded and directed a Division of Health Promotion at a major hospital system. Later he founded a company that involved developing a national computing network for interactive, automated healthcare recruiting. His research, teaching and writing have been directed toward improving the use of computers for patient education and toward enhancing provider's interpersonal skills. Currently, as President of Healthbridge Systems, he is consulting and focusing on the development of multimedia training to prevent workplace illness and injury.

Matthew W. Kreuter, PhD, MPH, is an assistant professor in Behavioral Science/Health Education, and Director of the Health Communications Research Laboratory at the Saint Louis University School of Public Health. He has developed and evaluated computer-tailored interventions to promote smoking cessation, dietary changes, physical activity, cancer screening, childhood immunizations, and the treatment of alcoholism among patients in health care settings. He teaches a graduate course in Health Communication, and has published a number of articles in journals such as *Health Psychology, Health Education Research: Theory and Practice, The Ameri-*

can *Journal of Preventive Medicine*, and *The Journal of General Internal Medicine*. Dr. Kreuter received his PhD and MPH in Health Behavior and Health Education from the University of North Carolina at Chapel Hill.

Debra A. Lieberman, PhD (Stanford University, 1986), is a research, evaluation, and instructional design consultant for educational multimedia. She was an assistant professor in the Department of Telecommunications at Indiana University, where her research and teaching focused on human–computer interaction, uses and effects of educational technology, and children's processes of learning with interactive media. Her work in consumer health has included research and development of health education software for Caresoft, InterPractice Systems, and Raya Systems. She has also worked with software companies, such as Computer Curriculum Corporation, Paramount Interactive, and Pixar, to conduct research and user testing, and to design children's multimedia products for the home and school.

Fiona McTavish, BA (Seattle Pacific University, 1980) is Assistant Researcher in the Center for Health Systems Research and Analysis. Her research interests include how people get information and how it contributes to their quality of life.

Betta Owens, MS (University of Wisconsin–Madison, 1992) is Research Program Manager in the Center for Health Systems Research and Analysis. Her research interests focus on issues that arise when introducing innovations to large organizations.

Suzanne Pingree, PhD (Stanford University, 1975) is Professor of Family and Consumer Communication at the University of Wisconsin–Madison. Her research interests include television and cognition, and using new technology in health communication.

Rajiv N. Rimal, PhD, is Assistant Professor in the Department of Speech Communication at Texas A & M University. His primary areas of interest include the adoption of multilevel approaches in health behavior change, research on the effectiveness of public health campaigns, and the application of new technology in health promotion. He is currently working on projects that investigate the role of family communication patterns on health behavior.

Dirk Scheerhorn, PhD (University of Iowa, 1989), is Associate Professor of Health Communication at the Philadelphia College of Pharmacy and

Science. His interests center on communication by patients and consumers and within illness-related groups. Dr. Scheerhorn's work has appeared in several journals and edited anthologies including *Health Communication, Communication Research,* and *Communicating About Communicable Disease.* He currently serves as vice-chair for information at the National Hemophilia Foundation (New York, NY).

Celette Sugg Skinner, PhD, is Assistant Professor, Mallinckrodt Institute, Washington University School of Medicine and a member of the Prevention & Control and Human Cancer Genetics Research Programs of the Washington University Cancer Center. Dr. Skinner's research examines the behavioral decisions involved in cancer screening and develops and evaluates behavior change interventions for cancer control. She has developed computer-tailored interventions targeting a number of health-related behaviors, using a variety of media. Her research has been published in various journals including *The American Journal of Public Health, Patient Education and Counseling, Health Education Research: Theory and Practice, Cancer Detection and Prevention*, and *Psychology, Health & Medicine.* She holds an MA in Communications Research from the Wheaton Graduate School and a PhD in Health Behavior and Health Education from the University of North Carolina School of Public Health.

John E. Wennberg, MD, MPH, is the Director of the Center for the Evaluative Clinical Sciences at the Dartmouth Medical School. He has been a Professor in the Department of Community and Family Medicine since 1980 and in the Department of Medicine since 1989 and currently holds the Peggy Y. Thomson Chair for the Evaluative Clinical Sciences. Dr. Wennberg is a member of the Institute of Medicine of the National Academy of Science. He has received a number of awards, including the Association for Health Services Research's Investigator Award, the Baxter Foundation's Health Services Research Prize, and the Quality Award from US Healthcare and USQA for outstanding contributions in the fields of small area analysis, outcomes research, and informed patient decision making.

Meg Wise, MS (University of Wisconsin–Madison, 1975), is Senior Research Specialist in the Center for Health Systems Research and Analysis. Her research interests include how computers and other communication technologies can help people adopt and sustain lifestyle changes to manage chronic illness.

I

THEORETICAL PERSPECTIVES

Health Promotion and Interactive Technology: A Conceptual Foundation

Richard L. Street, Jr.
Texas A&M University and Health Science Center
Rajiv N. Rimal
Texas A&M University

A middle-aged man has difficulty keeping his diabetes under control. He goes to the clinic where a nurse educator suggests he either attend a week-long diabetes education class or schedule an appointment with a diabetes educator after receiving individualized instruction using "Living with Diabetes," an interactive computer program. The multimedia program provides information on diabetes management and presents a series of problem-solving simulations (e.g., managing a hypoglycemic reaction, what to do on sick days). Because work and family commitments prevent his attendance at diabetes classes, he decides on the individualized instruction followed by an appointment with the nurse educator.

A woman is experiencing considerable stress and frustration in taking care of her elderly mother, who is suffering from Alzheimer's disease. Her mother's physician recommends that, at a modest cost, she rent a computer and modem so that she can access a computer network dedicated to care givers of people with Alzheimer's disease. By using the network, she would be able to leave messages for the clinical staff, receive a response within 24 hours, search a library for information on caring for the elderly, schedule clinical appointments, and discuss her experiences with other care givers in the network through a bulletin board and electronic mail service.

In the two scenarios described the use of interactive technology for health promotion seems appropriate and appealing. It is hard to think that a brochure, videotape, telephone, or even a professional consultation would be as versatile or as efficient a means of providing such a variety of health information and services. Empirically, however, the question is largely unanswered: What media and forms of communication are the most effective for health promotion and patient education initiatives? On a broader level, this is the classical question researchers investigating media effects have been asking for over half a century. So far, the answer seems to be that the media used in health promotion interventions affect some of the people some of the time under some conditions (Flora, Maibach, & Maccoby, 1989; Rogers & Storey, 1987). The aim of this book is to unpack these processes with particular attention to interactive technologies. As a starting point, we begin our discussion with a working definition of interactive technology.

INTERACTIVE TECHNOLOGY DEFINED

In this chapter and throughout this book, the term *interactive technology* is used to refer to computer-based media that enable users to access information and services of interest, control how the information is presented, and respond to information and messages in the mediated environment (e.g., answer questions, send a message, take action in a game, receive feedback or a response to previous actions). Two technological features of interactive media are *interactivity* and *modular components.*

The concept of interactivity has been defined in a variety of ways (Rafaeli, 1988). For our purposes, Biocca (1992) offered a useful definition: Interactivity refers to "(a) the number and forms of input and output, (b) the level of responsiveness to ... user actions and states, and (c) the range of interactive experiences (including applications) offered by the system" (p. 64). Two important capabilities created by interactivity are responsiveness and user control. *User control* refers to the extent to which the user of the system can participate in modifying the form and content of the mediated environment (Steuer, 1992). User control allows the user to determine what topics or services are selected, the order in which these selections are made, and the ways in which he or she can respond to information presented in the program or network. *Responsiveness,* on the other hand, refers to the extent to which a response takes into account the form, content, and nature of a previous action (Rafaeli, 1988). For example, a computer program that simply switches screens after a keystroke is more reactive than interactive. A more responsive program might generate a certain response (e.g., praise or corrective feedback) depending on previous information provided by the user (e.g., a correct or an incorrect answer to a question). A highly responsive program might be a game or simulation where the responses of the program are continually changing in form and content and are directly contingent on specific actions by the user (e.g., a person versus a computer chess game or a flight simulator).

As opposed to a single component (e.g., a brochure or videotape), interactive media are comprised of individual modular units that are linked together through programming. Modular components enable the program to utilize a diverse array of databases (animation, narration, graphics, text, motion pictures, music) within a single device or link together a variety of services (e.g., electronic mail, library data banks) within a single network. Thus, rather than making a single presentation in a predetermined linear sequence, interactive computing enables users to access different parts of the program and move with relative ease from one domain to another (Dede & Fontana, 1995).

HOW CAN INTERACTIVE TECHNOLOGY BE USED FOR HEALTH PROMOTION?

There are basically three types of computer applications pertinent to health promotion and patient education. First, many programs are developed to provide an information environment so that a user can learn about a particular health topic. The primary assumption here is the link between enhancement of knowledge and improvement in health. Although numerous studies and mass media campaigns have established this link (see, for example, Chaffe & Roser, 1986; Singer, Rogers, & Glassman, 1991), the unique aspect of interactive technology is its ability to promote active learning, information seeking, and individualized knowledge by allowing users to select information on topics of interest, access multiple modes of information (e.g., video demonstrations, stories, statistics), and direct their own path through the program (Dede & Fontana, 1995; Kahn, 1993). We use the term *information environment* to refer to programs of this sort because the user has a sense of presence within the media and can actively explore the information.

The primary goal of the program described in the chapter opening first scenario was to provide a self-directed tutorial enabling the user to acquire more knowledge about a health or medical issue—diabetes management in this case. In addition to learning, these types of programs may accomplish other objectives as well, such as reducing the user's anxiety about a condition or treatment (e.g., taking insulin injections) and motivating the user to enact certain behaviors (e.g., changing dietary habits). Represented in this book are interactive programs providing information environments pertaining to health crises (e.g., AIDS, breast cancer; chapter 5, this volume), asthma and diabetes in children (chapter 6, this volume), and patient involvement in medical decision-making (chapter 7 and 8, this volume).

Although much less utilized, interactive technology also can provide simulation environments for problem-solving and practicing disease-management skills. Compared to information environments, simulation environments are more reality-based in that they attempt to represent lifelike situations and experiences. A simulation environment might function like a video game where the user encounters various problematic situations or exigencies. For example, a program on diabetes management might simulate the beginning of a day where the person using the program could choose something to eat for breakfast, and then open a window to examine the food's effect on blood sugar. Then, placed in the context of driving to work, the user might unexpectedly have a flat tire that leads to stress and unexpected exertion. The program then signals that the user is having a hypoglycemic reaction. The user would then choose an action (e.g., drink fruit juice), see if this produces the desired effect, and then continue with the simulated day. With multimedia technology, these simulations can be produced in a very interesting, entertaining, and lifelike manner. Lieberman (chapter 6, this volume) and Manning (chapter 4, this volume) discuss the potential of simulation environments for helping users acquire a sense of self-efficacy and develop problem-solving skills.

Theoretically, the importance of a simulated environment in promoting health is based on its potential to act as a powerful source of self-efficacy (Bandura, 1977; Strecher, De Vellis, Becker, & Rosenstock, 1986) and to offer opportunities for practicing problem-solving skills (Dede & Fontana, 1995). Although performance accomplishments comprise the primary source of self-efficacy, simulations of real-world events can be created with interactive technologies that, although not real life, are nonetheless lifelike and thus provide a vicarious learning environment that resembles direct experience. Thus, simulation environments have the potential to personalize the experience, emphasize individual responsibility, and promote internalization of the knowledge and application of skills (Dede & Fontana, 1995). Of course, information and simulation environments can be integrated so that a user not only can learn about a health condition and its management, but also have opportunities to apply this knowledge to an array of hypothetical situations.

Yet another application of interactive technology is the creation of computer networks that enable a user to access (e.g., with a home computer and a modem) other people and computers in the network. By utilizing the network, the user can exchange messages with other people linked to the system, solicit medical advice and information from experts, and access medical libraries and other databases in the system or on the World Wide Web. The woman taking care of her elderly mother (described in the second scenario earlier) might use such a system to share her experiences with other care givers, leave a message for a health care provider, peruse an electronic bulletin board for announcements, access articles on care giving, and even schedule a clinical appointment.

The potential advantage of computer networks is that they offer a means for providing, within a single system, a variety of health-related resources including information, interaction with providers, social support, and decision support (Brennan, Moore & Smyth, 1995 Gustafson et al., 1993). In fact, although stand-alone programs currently offer the most interactive, most complex, and most sensory vivid information and simulation environments (e.g., on CD-ROM and hard drives), it is likely that networks will one day provide access to these multimedia environments as bandwidth technology improves and becomes more readily available in homes, offices, health clinics, and community centers. Using computer networks to deliver health services is a topic addressed in three chapters in this book (chapters 5, 9, and 10, this volume).

IS INTERACTIVE TECHNOLOGY
A FEASIBLE AND EFFECTIVE MEDIUM
FOR HEALTH PROMOTION?

A Comparison to Other Media

One way to answer this question is to identify media features that can enhance the effectiveness of health promotion initiatives and then examine how interactive technology fares on these dimensions, compared to more traditional methods of health promotion (e.g., professional consultation, videotape, printed material, telephone library). There are at least six features of health promotion media that may enhance their effectiveness—interactivity, networkability, sensory vividness, modifiability, availability, cost and ease of use.

Table 1.1 compares several types of media in relation to these features. As can be seen, interactive media fare well compared to other media. The potential advantages of interactive technology stem from its two defining

TABLE 1.1
A Comparative Analysis of Various Health Promotion Media

Criteria	Media				
	Interactive Technology	Professional Consultation	Telephone Systems[b]	Videotape	Brochure
Interactivity	high[a]	high	moderate	low	low
Sensory Vividness	high	high	low	high	low
Networkability	high	moderate	high	low	low
Modifiability	high	high	high	low	low
Availability	low	variable[c]	low	high	high
Cost	moderate	high	moderate	low	low
Easy to use	variable	variable	variable	variable	variable

[a]High, moderate, and low represent how the media tend to score on these dimensions. Variable indicates wide variation on how the medium scores on that dimension.

[b]By telephone systems we are referring both to use of the telephone to talk to others as well as to access recorded information in a phone databank.

[c]For example, people living in rural areas may have very limited access to health care providers.

characteristics: interactivity and modular components. Interactivity allows the user to act on the program to control its course, direction, and presentation of content. By so doing, the user utilizes the information or system in accordance with his or her individual needs and interests (Kahn, 1993; Skinner, Siegfried, Kegler, & Strecher, 1993). Interactivity also may facilitate involvement because the user actively participates in the experience by choosing services or topics to view, selecting videoclips to watch, repeating topics or services if desired, and responding to information in the system (Kozma, 1991). Some research has shown that interactivity enhances interest, active information processing, and satisfaction with a message (Dede & Fontana, 1995; Rafaeli, 1988; Schaffer & Hannafin, 1986), which in turn contributes to the effectiveness of persuasive and educational materials (Chaffe & Roser, 1986; Petty & Caccioppo, 1986; Webber, 1990).

Interactive media also can be used to create a computer network that enables the user to access databases (e.g., medical libraries, recorded experiences of others) and to communicate with others linked to the system (e.g., health care providers, others confronting similar health problems). As several chapters in this volume demonstrate, computer networks can offer a more efficient and effective means of providing access to social support, medical expertise, and assistance in making health-related decisions than do more traditional media such as telephones and face-to-face contact (chapters 5, 7, and 8, this volume).

The use of modular components in interactive media offers additional advantages. First, the components of the program can include multiple modalities of delivering information (e.g., text, narration, motion picture, graphics, music). A program that uses an array of media components creates a more vivid presentation because the receiver experiences more sensory stimulation (e.g., sight, sound, color, movement, reading) during message processing (Biocca, 1992; Steuer, 1992). The use of multiple channels also enables the program to accentuate and reinforce informational content (e.g., using both narration and motion picture to describe how to conduct breast self-examination). Sensory vividness and information enhancement appear to be two reasons why health promotion materials that present information through multiple modalities (e.g., computers, videotapes) are more effective than are materials that rely solely on a single channel (e.g., brochures; Gagliano, 1988; Marshall, Rothenberger, & Bunnell, 1984).

Second, because they are created with individual modular components, interactive programs are more modifiable than are other media. This is important because the needs of providers and clients may change over time and vary across locations, institutions, and populations. Programs with modular components can offer the user a choice of services and topics, they can be customized to accommodate the expertise and policy of a particular health care organization or network, and they can be updated to stay abreast of state-of-the-art knowledge or services. This is in contrast to the creation of a solidified presentation (e.g., videotape, brochures, public service announcement) that cannot be easily adapted to the needs of different users and has a set lifespan prior to its unavoidable obsolescence.

Admittedly, other media offer some of these advantages (see Table 1.1). For example, professional consultations are highly interactive, engage multiple senses, and (presumably) are capable of providing the most up-to-date knowledge and services. However, the interactions that occur during these encounters tend to be controlled by the health care professional and may circumvent the patient/client's individual's needs and interests (Street, 1992). Although phone banks of recorded messages are interactive (e.g., a touch-tone phone is used to select or leave messages), these media do not allow for the rapid transmission and exchange of information, mobility between different components of the program, diversity of data formats, and data storage that are possible with computers. Videotaped presentations and public service announcements use audiovisual channels to present messages, yet these media are largely noninteractive, as the viewer must sit and watch the presentation in a predetermined linear sequence (Steuer, 1992).

On some criteria, interactive technology appears to fare less favorably. First, there is the perception, especially among nonusers, that one needs special expertise to use a computer. As noted in Table 1.1, however, every form of health promotion has the potential to be user unfriendly. For example, many brochures and videotapes present information at a literacy level that exceeds that of the intended audience (Davis, Crouch, Wills, Miller, & Abdehou, 1990). Dialing into a phone databank can be extremely frustrating as one tries to listen to and remember all the options and codes prior to selecting one. Even the person-to-person interaction of a professional consultation may be fraught with problems of communication due to differences between the interactants' status, knowledge, and attitudes toward health and behavior (Street, 1991; Waitzkin, 1991). Of course, operating a computer can be difficult. However, interactive programs are currently being developed that are easy to use (e.g., touchscreen monitors) and suitable for (even preferred by) people of all ages and levels of education (Barry, Fowler, Mulley, Henderson, & Wennberg, 1995; Gustafson et al., 1994).

Finally, for a variety of reasons, health-care institutions, insurance companies, and public-health interests have been slow to invest in computing technology for health promotion and patient education. Computer systems and software are still relatively expensive compared to the other media (although less expensive than professional consultations), and currently only a limited number of programs are available. Nevertheless, as the technology becomes more affordable and there is greater selection of software programs, computers will, in the near future, be as common a health education resource as VCR players and videotapes are now. Furthermore, although interactive technologies may require higher up-front costs than other educational media, the long-term costs will be lower if use of these programs lead to healthier behavior, more knowledgeable patients, more appropriate use of clinical services, and fewer or shorter visits with health care providers.

Improving Health Outcomes

The shortage of interactive health programs and services will remain until public health and health care stakeholders (e.g., providers, patients, communities, businesses, insurance companies) are convinced that there are better outcomes when interactive media are used to deliver health promotion programs and services. Unfortunately, relatively few efforts have been made to articulate and critically examine theoretical frameworks that should guide the development and evaluation of these systems (Brennan, 1995;

chapter 2, this volume). Furthermore, relatively little research has compared the effectiveness of various media for delivering health messages and services.

We have listed in Table 1.2 a number of studies that have examined the effectiveness of using interactive media for health promotion, patient education, and the delivery of health care services. At first glance, these studies indicate that interactive technology is a very effective means of providing health resources. However, a closer inspection of the research suggests we may need to temper our enthusiasm. For example, the first group of studies listed in Table 1.2 either compared pre- and post-use of a particular application on an outcome of interest (e.g., knowledge) or compared outcome scores among users of the program to a control group that received no intervention. Although these studies suggest that better outcomes were associated with the use of interactive media, one does not know whether the results were due simply to the provision of the health resource per se (regardless of the media), or to some inherent feature of the technology and the user's experience within the mediated environment. The second group of studies also found better outcomes with the use of interactive technology. However, these studies examined the effectiveness of computer-assisted instruction *in addition to* other educational resources. Thus, improvement in outcomes may have been related to some unique feature of the user–media–message interaction or to the fact that the users of computer programs simply received additional or reinforcing information.

The final group of studies compared the effectiveness of different media to provide health information or services. Considered collectively, these investigations appear to indicate that interactive media are at times superior to, and at times no better than, other media with respect to educational and health outcomes (see also Jelovsek & Adebonojo, 1993). However, a global statement of this sort may be misleading because, in addition to differences in content and targeted audiences, there are important differences among the media environments used in these studies. For example, some of the computer programs used text only; others used multimedia formats. Some were interactive only to the extent that they allowed users to change screens with keystrokes (not much different from turning the pages of a pamphlet); others were interactive in that users could branch to different topics, add information to the program, and receive feedback on previous responses. In short, we will understand the value of interactive media only when researchers and developers can adequately explain how the attributes of user, message, and media interact in such a way as to affect knowledge, attitudes, and behavior within a specific health domain.

TABLE 1.2

Studies Evaluating the Effectiveness of Interactive Media (IM) and Computer Networks (CN) for Health Promotion

Authors	Topic	Design	Results	Interactive Media Better?
Group 1: Pre vs. post; computer versus control				
Alemi & Higley, 1995	Health risk assessment	IM[a] vs. Control	1. No difference in intent to reduce risk	o
Brennan et al., 1995	Caring for persons with Alzheimer's	CN vs. Control	1. Greater confidence in decision-making 2. No difference in social isolation or decision-making skill	+/o
Gustafson et al., 1994	People living AIDS/HIV	CN vs. Control	1. Heavy use of CN by experimental group 2. Improved quality of life 3. Reduction of health care costs	+
Kinzie, Schorling, & Siegel, 1993	Prenatal alcohol use	Pre- vs. Post-IM	1. Increase in knowledge 2. Increase in intention to reduce consumption	+
Lyons, Krasnowski, Greenstein, Maloney, & Tatorczuk, 1982	Heart disease and treatment	Pre- vs. Post-IM	1. Increase in knowledge	+
Rippey et al., 1987	Management of osteoarthritis	Pre- vs. Post- IM	1. Increase in knowledge 2. Increase in self-reported management behaviors	+
Robinson, 1989	Self-care and disease prevention	CN vs. Control	1. Decrease in doctor visits 2. Increased in self-efficacy to prevent STDs	+
Wagner et al., 1995	Treatment options for benign hypertrophy	Pre- vs. Post- IM	1. Decrease in surgery and increase in "watchful waiting"	+[b]
Wetstone, Sheehan, Votaw, Peterson, & Rothfield, 1985	Management of rheumatoid arthritis	IM vs. Control	1. Increase in knowledge 2. Increase in joint protection behavior 3. Improved outlook	+

(Continued)

Authors	Topic	Design	Results	Interactive Media Better?
Winett, Moore, & Wagner, 1991	Food purchases	IM vs. Control	1. Increased high fiber purchases 2. Decreased high fat purchases	+
Group 2: Computer program as supplement				
Ben-Said, Consoli, & Jean, 1994	Hypertension	Counseling + brochure vs. IM + counseling + written	1. Increased knowledge with both interventions 2. Larger knowledge gain with computer supplement	+
Huss et al., 1992	House dust mite avoidance	Counseling + written vs. IM + counseling + written	1. Decrease in mite allergens 2. Decrease in allergy symptoms	+
Group 3: Computer versus other media				
Alterman & Baughman, 1991	Alcohol treatment	IM vs. Video	1. Knowledge gain in both conditions 2. No difference between the two groups	o
Deardoff, 1986	Sexually transmitted disease	IM vs. Face to face vs. Written vs. Control	1. Increased information recall with computer and written 2. More positive evaluation of computer and face to face	+/o
Kritch et al., 1995	AIDS	High interactive IM^2 vs. Low interactive IM vs. Video	1. Users of high interactive program gained more knowledge than other groups	+
Kumar, Bostow, Schapira, & Kritch, 1993	Nutrition	High interactive IM^2 vs. Low interactive IM vs. Written	1. Users of high interactive program gained more knowledge and decreased fat intake more than other groups 2. Interactive program took longer to complete	+
Rickert et al., 1993	Alcohol education for adolescents	IM vs. Physician vs. Control	1. More knowledge gain with computer and physician than with control	o

(Continued)

11

Authors	Topic	Design	Results	Interactive Media Better?[b]
Rubin et al., 1986	Managing asthma (for children)	IM vs. Doctor/nurse	1. Increased knowledge and adherence with computer 2. Decrease in number of acute visits with computer	+
Schneider, Walter, & O'Donnell, 1990	Smoking cessation	Comprehensive CN vs. Limited CN[d]	1. Low abstinence by users in both groups 2. Users in comprehensive group more likely to stay in the program and have higher abstinence	+
Selmi, Klien, Griest, Sorrell, & Erdman, 1993	Depression therapy	IM vs. Therapist vs. Control	1. Decrease in depression with computer and therapist vs. control	o
Street, Voight, Geyer, Manning, & Swanson, 1995	Options for treating breast cancer	IM vs. Brochure	1. Increased knowledge with both interventions 2. Knowledge gains higher with computer 3. No difference in patient involvement in decision-making	+/o
Tibbles, Lewis, Reisine, Rippey, & Donald, 1992	Joint replacement surgery	IM vs. Doctor or nurse	1. Increase in knowledge with computer	+
Wise, Dowlashati, Farrant, Fromson, & Meadows, 1986	Diabetes management	Multimedia IM + knowledge test vs. Multimedia IM vs. Control	1. Increased knowledge and lower blood sugar with both IM groups compared to control 2. No significant differences between the two IM groups	+

[a]IM refers generally to the use of computing and interactive technology to deliver health information.

[b]+ indicates better outcomes with interactive media compared to other conditions. +/o indicates mixed results. o indicates no differences in outcomes due to type of media.

[c]High IM refers to programs that either (a) required users to provide some keywords and answers when using the program and/or (b) allowed users to review previous sections of the program. Low IM refers to programs where the user tapped a key to move the screen forward.

[d]The comprehensive computer network (CN) had access to a stop smoking forum, communication with others, and personalized behavioral program. The limited CN had was much shorter and provided fewer services.

A CONCEPTUAL FRAMEWORK

Figure 1.1 presents a three-stage model of health promotion using interactive technology. The model is not intended to be a comprehensive, explanatory theory. Rather, its purpose is to serve as an organizing and heuristic framework that identifies specific variables and processes that will determine the effectiveness of interactive media environments designed to enhance the health and well-being of a particular audience.

Stage 1 Processes: Implementation and Utilization

The question of whether interactive technology will help the adult diabetic or the woman caring for her elderly mother is moot if these programs are not available and, even if available, are not used by the targeted audience. Stage 1 focuses on how the interplay of institutions (e.g., health care organizations, work sites, public health agencies), users (e.g., patients, employees at a worksite, special populations), and the technology (e.g., internet, CD-ROM, WANs; etc.) will affect implementation and utilization of interactive technology for health promotion purposes.

At the institutional level, some of the important variables include organizational inertia toward change, providers' attitudes toward patient education and preventative care, the existence of computing facilities at the

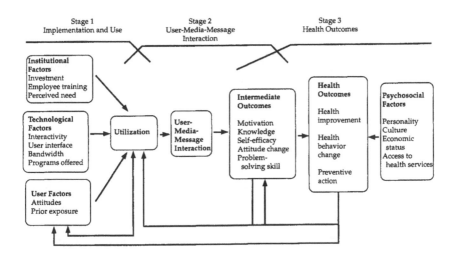

FIG. 1.1.   A three-stage model of health promotion using interactive technology.

worksite or clinic, and whether there is a culture of computer use within the organization. At the user level, one's predispositions to use interactive technology may be related to sociodemographic variables (e.g., gender, education, age), attitudes toward computers, familiarity with the technology, self-efficacy regarding one's computer literacy, and whether computers are available at home or work (Dutton, Rogers, & Jun, 1987). Finally, variables pertaining to the technology itself include cost and availability, bandwidth, cable lines, and the number and quality of educational, self-care, and support programs available.

Although our primary emphasis in this book is on the user–media–message interaction and health outcomes resulting from these experiences (Stages 2 and 3 in Fig. 1.1), three chapters (chapters 11, 12, and 13) in this volume address some of the policy, technological, economic, institutional, and user issues regarding the future of health promotion using interactive technology. (See also Dutton et al., 1987; Fisher, 1995, Gorry, Harris, Silva, & Eaglin, 1995, and Negroponte, 1995).

Stage 2 Processes: User–Media–Message Interaction

A key premise in this book is that the effectiveness of health promotion interventions depends, in large part, on how users process health messages (e.g., engagement, attention, integration of information with existing knowledge) and whether this experience produced the desired results (e.g., learning, motivation, enjoyment, problem-solving skill, reassurance). How message processing unfolds will depend on the interaction of user, media, and message characteristics. Important variables associated with users include the perceived relevance of the topic or need for the service, desire for information, emotional state, and attitudes toward the medium used to present the health message or service. Characteristics of the media include ease of use, degree of interactivity, sensory vividness, and the speed with which one can move through different parts of the system. Message characteristics include the health topic or service, informational content, reasoning and evidence provided, and the credibility of message sources (as perceived by the users).

Message processing itself may be analyzed along a variety of dimensions including the user's cognitive involvement in the message (e.g., perceived importance, attention to content), affective responses (e.g., excitement, apprehension, frustration, enjoyment) to the message and the technology, and the user's path through the environment (what topics or services are selected, in what order, for how long a period of time). Generally, it is hoped that the immediate outcomes of this experience would be a better understanding of a health issue, more resources for managing a health situation (e.g., emotional support, disease management strategies), an enhanced sense of self-efficacy, and more motivation to engage in the targeted health behaviors (see Fig. 1.1).

Important questions related specifically to the use of media include: Does one learn differently when using interactive technology than when reading a pamphlet, talking to a doctor, participating in a discussion group, or watching a videotape? What features of the technology are most conducive to stimulating involvement in message processing? Do users differ in their preferences for educational media? Are these preferences affected by the user's learning or coping style (for example, information seekers versus information avoiders)? Answers to these questions will shed light on how the technology might be developed and adapted to promote better outcomes (see also Kozma, 1991).

Stage 3 Processes: Health Behaviors and Outcomes

The goal of a health promotion intervention typically is to provide messages and/or services that are sufficiently powerful (e.g., informative, persuasive, useful) to make a lasting contribution to the user's ability to maintain and improve health and to make appropriate and satisfying health-related decisions. This is an ambitious task, given that health-related beliefs and behaviors also are affected by a milieu of psychological, social, cultural, and economic processes. To be effective, health promotion programs and services must be designed and delivered in a manner responsive to these contextual exigencies. For example, after receiving breast cancer education, a woman may intend to schedule a mammogram. Yet, this intent may dissipate if she cannot afford the service or has no convenient means of scheduling the appointment. Similarly, the man with diabetes (described earlier) may be motivated to change elements of his lifestyle (e.g., diet, exercise) following diabetes education, but the motivation may quickly wane once he gets away from the mediated environment and back to his life world of everyday demands, family eating habits, stress at work, and so on.

In short, one cannot rely solely on the computer program or network to singlehandedly accomplish health promotion objectives. Interactive technology can be an extremely valuable resource but, like other forms of health promotion, it must be responsive to and integrated within the various social, cultural, and economic environments affecting health behaviors and outcomes of interest. Health promotion planners, educators, and researchers will need to address these exigencies (e.g., with social support, follow-up assessment, continuing education) in order to enhance the effectiveness of the intervention.

CONCLUSION

In summary, the effectiveness of health promotion and patient education using interactive technology depends on a host of processes that operate at many different levels, ranging from institutional policies to the individual user's life world following use of the technology. It is important to note that we depict these processes as cyclical and not linear phenomena (see Fig. 1.1). Obviously, interactive technology will have little effect on health outcomes until there are more programs available and in use by health care institutions, patients, consumers, work sites, and other stakeholders. On the other hand, the success of these programs (Stage 3) will in turn contribute to the likelihood that health care providers will implement the systems (Stage 1) and that users will find interactive media as an acceptable, if not valuable, method of receiving health information and services (Stage 2).

REFERENCES

Alemi, F., & Higley, P. (1995). Reaction to "talking" computers assessing health risks. *Medical Care, 33,* 227–233.

Alterman, A. I., & Baughman, T. G. (1991). Videotape vs. computer interactive education in alcoholic and nonalcoholic controls. *Alcoholism: Clinical and Experimental Research, 15,* 39–44.

Bandura, A. (1977). Self-efficacy: Toward a unifying theory of behavioral change. *Psychological Review, 84,* 191–215.

Barry, M. J., Fowler, F. J., Mulley, A. G., Jr., Henderson, J., Jr., & Wennberg, J. E. (1995). Patient reactions to a program designed to facilitate patient participation in treatment decisions for Benign Prostatic Hyperplasia. *Medical Care, 33,* 771–782.

Ben-Said, M., Consoli, S., & Jean, J. (1994). A comparative study between a computer-aided education (ISIS) and habitual education techniques for hypertensive patients. In J. Ozbolt (Ed.), *Proceedings of the Annual Symposium on Computer Applications in Medical Care* (pp. 10–14). Philadelphia: Hanley & Belfus.

Biocca, F. (1992). Virtual reality technology: A tutorial. *Journal of Communication, 42, 23–72.*

Brennan, P. F. (1995). Characterizing the use of health care services delivered via computer networks. *Journal of the American Medical Informatics Association, 2,* 160–168.

Brennan, P. F., Moore, S. M., & Smyth, K. A. (1995). The effects of a special computer network on care givers of persons with Alzheimer's disease. *Nursing Research, 44,* 166–172.

Chaffe, S. H., & Roser, C. (1986). Involvement and the consistency of knowledge, attitudes, and behavior. *Communication Research, 13,* 373–399.

Davis, T. C., Crouch, M. A., Wills, G., Miller, S., & Abdehou, D. M. (1990). The gap between patient reading comprehension and the readability of patient education materials. *Journal of Family Practice, 31,* 533–538.

Deardoff, W. W. (1986). Computerized health education: A comparison with traditional formats. *Health Education Quarterly, 13,* 61–72.

Dede, C., & Fontana, L. (1995). Transforming health education via new media. In L. Harris (Ed.), *Health and the new media: Technologies transforming personal and public health* (pp. 163–184). Mahwah, NJ: Lawrence Erlbaum Associates.

Dutton, W. H., Rogers, E. M., & Jun, S. (1987). Diffusion and social impacts of personal computers. *Communication Research, 14,* 219–250.

Fisher, F. D. (1995). But will the new health media be forthcoming? In L. Harris (Ed.), *Health and the new media: Technologies transforming personal and public health* (pp. 209–227). Mahwah, NJ: Lawrence Erlbaum Associates.

Flora, J. A., Maibach, E. W., & Maccoby, N. (1989). The role of media across four levels of health promotion intervention. *Annual Review of Public Health, 10,* 181–201.

Gagliano, M. (1988). A literature review of the efficacy of video in patient education. *Journal of Medical Education, 63,* 785–792.

Gorry, G. A., Harris, L. M., Silva, J., & Eaglin, J. (1995). Health care as teamwork: The internet collaboratory. In L. Harris (Ed.), *Health and the new media: Technologies transforming personal and public health* (pp. 87–108). Mahwah, NJ: Lawrence Erlbaum Associates.

Gustafson, D. H., Wise, M., McTavish, F., Taylor, J. O., Wolberg, W., Stewart, J., Smalley, R. V., & Bosworth, K. (1993). Development and pilot evaluation of a computer-based support system for women with breast cancer. *Journal of Psychosocial Oncology, 11,* 69–93.

Gustafson, D. H., Hawkins, R. P., Boberg, E. W., Bricker, E., Pingree, S., & Chan C. (1994). The use and impact of a computer-based support system for people living with AIDS and HIV infection. In J. Ozbolt (Ed.), *Proceedings of the Annual Symposium on Computer Applications in Medical Care* (pp. 604–608). Philadelphia: Hanley & Belfus.

Huss, K., Squire, E. N., Jr., Carpenter, G. B., Smith, L. J., Huss, R. W., Salata, K., Salerno, M., Agostinelli, D., & Hershey, J. (1992). Effective education of adults with asthma who are allergic to dust mites. *Journal of Allergy and Clinical Immunology, 89,* 836–843.

Jelovsek, F. R., & Adebonojo, L. (1993). Learning principles applied to computer-assisted instruction. *M.D. Computing, 10,* 165–172.

Kahn, G. (1993). Computer-based patient education: A progress report. *M.D. Computing, 10,* 93–99.

Kinzie, M. B., Schorling, J. B., & Siegel, M. (1993). Prenatal alcohol education for low income women with interactive multimedia. *Patient Education and Counseling, 21,* 51–60.

Kozma, R. B. (1991). Learning with media. *Review of Educational Research, 61,* 179–211.

Kritch, K. M., Bostow, D. E., & Dedrick, R. F. (1995). Level of interactivity of videodisk instruction on college students' recall of AIDS information. *Journal of Applied Behavior Analysis, 28,* 85–86.

Kumar, N. B., Bostow, D. E., Schapira, D. V., & Kritch, K. M. (1993). Efficacy of interactive, automated programmed instruction on nutrition education for cancer prevention. *Journal of Cancer Education, 8,* 203–211.

Lyons, C., Krasnowski, J., Greenstein, A., Maloney, D., & Tatarczuk, J. (1982). Interactive computerized patient education. *Heart and Lung, 11,* 340–341.

Marshall, W. R., Rothenberger, L. A., & Bunnell, S. L. (1984). The efficacy of personalized audiovisual patient-education materials. *The Journal of Family Practice, 19,* 659–663.

Negroponte, N. (1995). *Being digital.* New York: Knopf.

Petty, R. E., & Cacioppo J. T. (1986) *Communication and persuasion: Central and peripheral routes to attitude change.* New York: Springer-Verlag.

Rafaeli, S. (1988). Interactivity: From new media to communication. In R. P. Hawkins, J. M. Wiemann, & S. Pingree (Eds.), *Advancing communication science: Merging mass and interpersonal processes* (pp. 110–134). Newbury Park, CA: Sage.

Rickert, V. I., Graham, C. J., Fisher, R., Gottlieb, A., Trosclair, A., & Jay, M. S. (1993). A comparison of methods for alcohol and marijuana anticipatory guidance with adolescents. *Journal of Adolescent Health, 14,* 225–230.

Rippey, R. M., Bill, D., Abeles, M., Day, J., Downing, D. S., Pfeiffer, C. A., Thal, S. E., & Wetstone, S. L. (1987). Computer-based patient education for older persons with osteoarthritis. *Arthritis and Rheumatism, 30,* 932–936.

Robinson, T. N. (1989). Community health behavior change through computer network health promotion: Preliminary findings from Stanford Health-Net. *Computer Methods and Programs in Biomedicine, 30,* 137–144.

Rogers, E. M., & Storey, J. D. (1987). Communication campaigns. In C. Berger & S. Chaffe (Eds.), *Handbook of communication science* (pp. 817–846). Beverly Hills, CA: Sage.

Rubin, D. H., Leventhal, J. M., Sadock, R. T., Letovsky, E., Schottland, P., Clemente, I., & McCarthy, P. (1986). Educational intervention by computer in childhood asthma: A randomized clinical trial testing the use of a new teaching intervention in childhood asthma. *Pediatrics, 77,* 1–10.

Schaffer, L. C., & Hannafin, M. J. (1986). The effects of progressive interactivity on learning from interactive video. *ECTJ, 34,* 89–96.

Schneider, S. J., Walter, R., & O'Donnell, R. (1990). Computerized communication as a medium for behavioral smoking cessation treatment: Controlled evaluation. *Computers in Human Behavior, 6,* 141–151.

Selmi, P. M., Klien, M. H., Griest, J. H., Sorrell, S. P., & Erdman, H. P. (1993). Computer-administered therapy for depression. *M. D. Computing, 8,* 98–102.

Singer, E., Rogers, T. F., & Glassman, M. B. (1991). Public opinion about AIDS before and after the 1988 U.S. Government public information campaign. *Public Opinion Quarterly, 55,* 161–179.

Skinner, C. S., Siegfried, J. C., Kegler, M. C., & Strecher, V. J. (1993). The potential of computers in patient education. *Patient Education and Counseling, 22,* 27–34.

Steuer, J. (1992). Defining virtual reality: Dimensions determining telepresence. *Journal of Communication, 42,* 73–93.

Strecher, V. J., De Vellis, B. M., Becker, M. H., & Rosenstock, I. M. (1986). The role of self-efficacy in achieving health behavior change. *Health Education Quarterly, 13,* 73–91.

Street, R. L., Jr. (1992). Communicative styles and adaptations in physician-parent consultations. *Social Science and Medicine, 34,* 1155–1163.

Street, R. L., Jr. (1991). Accommodation in medical consultations. In H. Giles, N. Coupland, & J. Coupland (Eds.), *Contexts of accommodation: Developments in applied sociolinguistics* (pp. 131–156). Cambridge, England: Cambridge University Press.

Street, R. L., Jr., Voigt, B., Geyer, C., Manning, T., & Swanson, G. (1995). Increasing patient involvement in deciding treatment for early breast cancer. *Cancer, 76,* 2275–2285.

Tibbles, L., Lewis, C., Reisine, S., Rippey, R., & Donald, M. (1992). Computer assisted instruction for preoperative and postoperative patient education in joint replacement surgery. *Computers in Nursing, 10,* 208–212.

Wagner, E. H., Barret, P., Barry, M. J., Barlow, W., & Fowler, F. J. (1995). The effect of a shared decision-making program on rates of surgery for Benign Prostatic Hyperplasia: Pilot results. *Medical Care, 33,* 765–770.

Waitzkin, H. (1991). *The politics of medical encounters.* New Haven, CT: Yale University Press.

Webber G. C. (1990). Patient education: A review of the issues. *Medical Care, 28,* 1089–1103.

Wetstone, S. L., Sheehan, T. J., Votaw, R. G., Peterson, M. G., & Rothfield, N. (1985). Evaluation of a computer based education lesson for patients with rheumatoid arthritis. *Journal of Rheumatology, 12,* 907–912.

Winnet, R. A., Moore, J. F., & Wagner, J. L. (1991). Altering shoppers' supermarket purchases to fit nutritional guidelines; An interactive information system. *Journal of Applied Behavioral Analysis, 24,* 95–105.

Wise, P. H., Dowlatshahi, D. C., Farrant, S., Fromson, S. & Meadows, K. A. (1986). Effects of computer-based learning on diabetes knowledge and control. *Diabetes Care, 9,* 504–508.

2

Interactive Technology Attributes in Health Promotion: Practical and Theoretical Issues

Rajiv N. Rimal
Texas A&M University

June A. Flora
Stanford University

The application of interactive technologies in health promotion efforts holds many promises, as chapters in this book document. To maximize the overall effectiveness of such efforts, theoretical development in this area has to stay ahead of the curve by developing models that are able to predict which technologies are best suited for which kinds of health promotion efforts. Although it is impossible to know the exact nature of technologies that will emerge as we enter the 21st century, we can, however, refer to the current literature about what humans do with and how they react to specific attributes of technology to make predictions about their potential use in health promotion. As with other media, research in this area will also be concerned with the fundamental question: Who uses them, for what purpose, and with what effect (see, for example, Blumler & Katz's, 1974, uses and gratifications approach)?

In this chapter, we propose a research agenda for synthesizing work and generating additional research questions. Our purpose is threefold: to summarize key aspects about the relationship among technology attributes, health behaviors, and individual characteristics; to present this summary with a framework that organizes knowledge in a meaningful way; and to propose future research that builds on extant knowledge about human–technology interactions.

Our research agenda proposes that we explore meaningful units of study for understanding the effectiveness of the various attributes of interactive technologies in health promotion. This approach will help define the boundary conditions for the pertinent theory, appropriate research questions, and expected outcomes. Our primary assumption is that more important than the technology itself is the configuration of attributes that comprise the technology. Not all attributes of interactive technologies will be used simultaneously or with the same frequency. Some attributes will be more effective for some purposes than for others, and some individuals will be more inclined to utilize some attributes and not others. Hence, we propose the study of interactive technologies according to the technology–health–users (THUS) framework shown in Table 2.1.

We return to this framework in the latter part of this chapter. Given the purpose of this book as a whole—interactive computing—and because a considerable body of knowledge already exists on the interactions among entries in the second and third columns, our primary focus will be on the first column, technology attributes.

ATTRIBUTES AND IMPLICATIONS OF INTERACTIVE TECHNOLOGIES

Although interactive technologies share many properties with other traditional media, their comparative advantage lies in their ability to combine

TABLE 2.1
The Role of Interactive Technologies in Health Promotion:
Research Using the Technologies–Health–Users (THUS) Framework

Technology Attributes	Health Domain Attributes	Attributes of Individuals
Multimodality	Addictiveness	Demographics:
Networkability	Duration of health benefits	Gender, age, education, income
Temporal Flexibility	Feasibility of self-help	Psychographics
Segmentation Capability	Maintenance of change	Knowledge, attitude, self-efficacy, perceptions, involvement
Interactivity	Knowledge requirement	Siociocultural aspects
Sensory vividness[a]	Skills requirement	Marital status, employment
Modifiability[a]	Importance of social support	Environmental aspects:
Availability[a]	Role of heredity	Extent of social networks
Cost[a]	Confidentiality requirement	
Ease of use[a]		

[a]Not discussed in this chapter; see chapter 1, this volume.

important features from various media into an integrated unit. This integration offers unique advantages and opportunities for health promotion. In the sections that follow, we discuss key attributes of interactive technologies that are expected to make a significant impact on health promotion efforts. The list of attributes we present is not meant to be exhaustive; it is only a sample of important features that, based on our current understanding, are expected to make a significant impact on health promotion efforts.

Multimodality

Integration of multimodality, such as text, pictures, video, and various combinations, into a single modular unit is a unique feature of interactive technologies. This integration allows for *multimodal versatility*. Compared to other media (e.g., television, newspapers, etc.), interactive technologies not only comprise the greatest number of modalities, but they also provide the flexibility to toggle between modalities and scale their relative salience. Hence, apart from the fact that they can evoke many similar psychological responses as a book, television, or videodisc, they have the additional capability of enhancing or decreasing the prominence of visual or auditory stimuli. Size of pictures, relative to text, for example, can be varied, as can the presence or intensity of sound.

Size of normal text in newspapers, a common source of frustration for older persons (O'Leary, Mann, & Perkash, 1991), can be enlarged for readability. Similarly, presentation of content through non-text-based formats can allow access by a wider range of users, particularly those underserved by conventional health promotion channels. This property of interactive technologies that allows for the manipulation of both formal features and content has important implications for learning, as we describe subsequently.

Multimodality also allows users to toggle between modalities and to change one to another without altering the original properties. Although this capability remains largely underutilized, it is possible, for example, to transform keyboard commands directly into sound (a property that has tremendous potential for the blind), or to input pictures, digitize, and subsequently alter them—all without changing the original stimulus.

An important contribution multimodality can make is the formation of simulated environments, also referred to as telepresence, "the experience of presence in an environment by means of a communication medium" (Steuer, 1992, p. 76), which evokes involuntary sensory experiences of being there (Reeves, 1991; Sheridan, 1992), or of "immersion" (Shapiro & McDonald, 1992, p. 106) in a mediated environment. These simulated environments can encourage role-playing and vicarious performances in various health contingency situations, such as a hypoglycemic attack (see the example in chapter 1).

Another important implication of multimodal capabilities of interactive technologies is demonstrated by research on how individuals with different levels of prior knowledge differ in how they process information presented in a multimodal format. Those who already possess a sufficient knowledge base tend to look at pictures only initially, whereas those without an adequate knowledge base tend to refer to pictures throughout the reading process (Rusted & Coltheart, 1979; Stone & Glock, 1981). It could be hypothesized that the former group refers to pictures initially to invoke relevant mental models. Once these models are activated, new information is simply deposited onto the same framework and the learning process proceeds fairly efficiently. For the latter group, on the other hand, lacking the requisite mental models, the learning process involves two processes: creation of new models and assimilation of new information into these models. This explains both the necessity of referring regularly to the pictorial model as text is being processed (Kozma, 1991), as well as lower rates of learning.

This implies that presentation of new information should be supplemented by external aides (Ross & Rakow, 1981) that facilitate the formation of mental models, as has been suggested elsewhere (Bates, 1992; Biocca, 1992). One such aide is redundancy across modalities. To the extent that text-based content is supplemented by pictorial models, for example, readers' abilities to learn will be greatly augmented. Learning and mastery of complex behaviors are enhanced by repetition and decomposition into simpler chunks (Bandura, 1977). Repetition and decomposition are further facilitated by redundancy of the message, and so the more redundant the message, the higher the likelihood of learning. However, redundancy is at odds with the novelty of presented information (Shannon & Weaver, 1949), and the latter is directly associated with attention paid to the message (Singer, 1980; Singer & Singer, 1979). Hence, redundant messages are less likely to elicit and hold viewers' attention. In the absence of sufficient amounts of attention, learning is unlikely to occur (McGuire, 1989).

Results from television research bear this out. Singer (1980) pointed out, for example, that because *Sesame Street*, relative to *Mr. Roger's Neighborhood*, uses more formal features that elicit children's attention (short edits, constant motion, and special sound effects), viewers have little time to adjust to events on the screen before being bombarded by newer ones. That is, redundancy levels are low and novelty of information is high. Because of attentional inertia, the time lag required for cognitive processing (Anderson, Alwitt, Lorch, & Levin, 1979), rates of learning are typically lower.

For learning to occur, a balance needs to be struck between redundancy levels and novelty of information. Multimodality can facilitate this process. pictures and sound can present material in novel ways through dramatization, role-playing, or simulation of sensory-rich environments (see chapter 1) to elicit the user's attention initially. Once attention has been elicited, opportunities then open up for presenting cognitively demanding material.

For some users, selection of modality may be based on necessity; because of the inability to overcome language barriers, for example, users may opt to view pictures rather than read text. Programs currently in the market already make this possible by presenting stories simultaneously on three modalities: text, sound that reads text on the screen, and pictures. Words being read aloud by the narrator can be highlighted on the screen. The built-in redundancy across visual and auditory modalities that makes this toggling possible has so far been limited primarily to children's programming and closed-captioning of television programs for the hearing impaired. The potential use of this feature of interactive technology for other target audiences remains largely untapped. Furthermore, because of current technological limitations and memory restrictions associated with depiction of motion on the computer screen (e.g., narrow bandwidth and slow CPU speeds), visual depiction of actual health behaviors has not yet become a widespread public-health practice.

Multimodality can also promote behavioral learning. Learning of behaviors occurs either through practice or by observing successful models enacting the behavior (Bandura, 1986). Although many behaviors in life are acquired by trial and error, through repeated performances that lead to subsequent mastery, others are learned through vicarious performance. As noted by Bandura (1977), it would indeed be grossly inefficient, if not impossible, to master all behaviors through actual performance. Instead, we observe what others do and evaluate the rewards or punishments they incur (Perry, Baranowski, & Parcel, 1990).

Social cognitive theory (Bandura, 1977) posits that the acquisition of behavioral skills is also related to model characteristics. Attractive models slightly higher in social prestige than the observer are much more likely to be emulated than those either lacking or much superior in social prestige. And, models perceived similar (in demographics, for example) are likely to be more successful in enhancing people's self-efficacy to enact the same behavior.

Because of the multimodal capabilities of interactive technologies, not only can successful models be depicted in an interesting and realistic manner, but, with proper design, model characteristics can be manipulated and matched with those of the user for optimal effect. Innovative attempts may even include presenting multiple models with diverse backgrounds and abilities and allowing the user to select one in a simulated role-playing environment. Although the efficacy of this dual strategy (matching model characteristics and providing choices to users) remains to be empirically tested, we would expect such efforts to yield positive results.

Networkability

Networkability refers to connections that computer users are now able to make with other users and service providers. The proliferation of Internet-access providers in the last few years and the subsequent decline in prices for their services have increased the number of online subscribers beyond most people's expectations (Chamberlain, 1994). America OnLine (AOL) alone, an Internet access provider, is expected to reach 9,000,000 American subscribers in 1998 (*The Economist*, 1996). Although the exact number of Internet users is difficult to ascertain, estimates range from about 20 to 30 million current users, a rate that is doubling every year (Elmer-Dewitt, Jackson, & Ratan, 1994). *The World Almanac* (1996) even lists a 341,000% annual rate of growth of World Wide Web traffic in 1995!

Because of this unprecedented growth of the online world, it has now become possible to communicate with others regardless of geographic distances, stay in touch with the latest innovations in science and medicine, and acquire specialized information with relative ease. Coupled with the passage of the Telecommunications Act of 1996, this growth, by most estimates, will greatly increase the potential role of computers in health care delivery, management, and education (Crawford & HIAWG, 1996).

The rise in computer networks is expected to impact both providers as well as receivers of health care. Managed care organizations (MCOs) can greatly enhance their reach among the currently underserved populations through telemedicine (Office of Rural Health Policy, 1994), which "uses a broad array of telecommunications technologies to bring health care to patients who are at a distance from the provider" (Crawford & HIAGW, 1996, p. 7). Providers throughout an integrated system can access and update patients' health records. Specialists in urban areas, for example, can remotely review rural patients' medical progress, thus providing essential services to an otherwise underserved population. The growth of teleradiology is a good example: in 1994, an estimated annual workload of five full-time radiologists (about 50,000 radiology procedures) was transmitted between U.S. facilities (Allen, 1994), and such procedures are estimated to grow 50% per year for the next 5 years (Crawford & HIAGW, 1996).

As computers make major inroads into U.S. homes, concerns about the disparity in access across ethnic lines are also mounting (Sutton, 1991). According to a recent Census Bureau study, for example, almost three times as many 3 to 17-year-old Whites than Blacks or Latinos had access to computers in 1993. Hence, in the absence of strong measures to ensure minority access, health information disseminated through computer channels is likely to miss those who need it the most. There are, however, significant changes in the horizon. In late 1994, for example, LatinoNet was founded by several nonprofit organizations to link community groups nationwide via America OnLine (Minority Markets ALERT, 1995). Similarly, Chicano/LatinoNet at the University of California at Los Angeles, through national and international links, provides free information for the Hispanic community.

Through networkability, health consumers will also be able to benefit from providers, other patients, or other caregivers. From providers, they can receive updates on their progress, as well as information about the latest technological breakthroughs in medicine. Networkability further provides opportunities for receiving social support from others undergoing similar health conditions (see, for example, Brennan, Moore, & Smyth, 1995) through either the use of chat groups (Ferguson, 1996), or an integrated online network (Gustafson et al., 1993). Computers are helpful in this regard particularly when the health condition is of such low prevalence in the population that geographical distances would make social support difficult to find. Similarly, patients are able to receive information about others' experiences with medical options and consequences (see Scheerhorn, chapter 10, this volume).

While advantages associated with networkability are numerous, perhaps the most consequential is the blurring of the distinction between source and receiver of messages. With interactive technologies, each receiver can also act as a sender (and vice versa). No longer do users have to rely on others' tastes, schedules, and predilections for selection of topic, modality, and content features, as is the case with traditional media. Once connected to a network, access to information from medical service providers, scientific publications, libraries, and online chat groups becomes virtually limitless.

Access to more information also means, however, that users have to assume a greater gatekeeping role in managing new information. In the absence of sound-filtering mechanisms, the flood of new information is likely to be more overwhelming than useful. Although it is increasingly becoming a standard business practice to set up home pages on the World Wide Web, for example, users who access those pages are greeted more often by company public relations material than by useful information. As with media literacy efforts that teach viewers to assume a more critical role (see, for example, Brookfield, 1986), computer literacy efforts are also required to help individuals, especially older adults whose prior experience with computers are minimal (Negroponte, 1995), navigate the complexities behind the national information infrastructure.

Once users of interactive technologies are connected to computer networks, feedback (defined broadly as the input users receive based on actions they perform) can be achieved on a regular basis. Feedback, in turn, is related to behavior because it is a powerful source of self-efficacy (Ross, 1995; Wallace & Alden, 1995). Positive outcomes, for example, can increase individuals' confidence in their own abilities, which increases the likelihood of further progress, especially if healthy attributional practices are invoked. Conversely, negative results can also act as sources of feedback, informing the individual that alternative methods may need to be explored. Furthermore, feedback is likely to lead to positive outcomes if it is construed as an incentive. Providing feedback to geographically isolated individuals, for example, can act as a form of social support.

Individuals can transmit their progress (a list of foods eaten every day, for example) to a central network, where experts (or, depending on the circumstances, even a computer) can make assessments and provide suggestions for future course of action. Feedback helps individuals evaluate their mastery at each stage before proceeding to the next, thus providing opportunities for corrective actions. When the desired healthy behavior is a difficult one, such as changing dietary habits, without corrective actions, either unhealthy or ineffective measures are likely to be practiced, incorporated, and learned, or, in the absence of positive outcomes, relapse is likely to occur.

With networked interactive technologies, not only is it possible to provide feedback, but the quality of the feedback itself can be enhanced by utilizing the unique features of interactive technologies. One such technique is to take advantage of the multimodal capability to provide feedback in a meaningful, interesting, and comprehensible manner. CHESS (Comprehensive Health Enhancement Support System) in the University of Wisconsin is an excellent example of how feedback and various other attributes of interactive technologies can be used successfully to provide information and social support to people living with HIV/AIDS infection (Boberg et al., 1995; Pingree et al., 1993).

Interactivity

Defined in chapter 1, *interactivity*[1] comprises two dimensions: responsiveness and user control. Whereas *feedback* is often used interchangably with *interactivity* (see, for example, Newhagen, Cordes, & Levy, 1995), for health promotion efforts, we find it more useful to conceptualize interactivity as a particular case of feedback that occurs in real time. In other words, interactivity cannot occur without feedback, but feedback can occur independent of interactivity, depending mostly on the amount of time required for the dual exchange of information. The greater the amount of time required for contingent responses to user actions, the less the interactivity of the system. Hence, a computer network that links users with health information providers or other users may serve as an excellent source of feedback (CHESS, as referred to earlier, for example) because experts or other users can transmit their assessments and suggestions back to the user. For such network systems to be truly interactive, they must provide feedback in real time.

Because interactivity subsumes feedback, feedback implications of interactive technologies discussed earlier also apply to interactivity, and will not be repeated here. An additional implication, however, has to do with perceived responsiveness. Interactions that occur in real time are perceived as more responsive and more interactive than time-delayed ones (Walther, Anderson, & Park, 1994; Rafaeli, 1988). Perceived responsiveness of the communication channel is correlated with heightened participation (Newhagen, 1994) and better performance (Watts, Baddeley, & Williams, 1982). Thus, interactive technologies responsive to users' actions are likely to be associated with positive outcomes.

Temporal Flexibility

Interactive technologies possess *temporal flexibility*, also referred to as *asynchroneity* or *asynchronicity* (Chamberlain, 1994; Rogers, 1986). Compared to traditional media, interactive technologies allow users to interact with the content as the content is made available, or to postpone the interaction as desired. Other existing technologies, such as the answering machine, already make the postponement of interaction possible. What is unique about interactive technologies is the myriad of possibilities they offer the user once the message has been retained. It is now up to the user to determine the circumstances and situational contexts within which the message will be processed. The user is also able to save the message, manipulate its content, and resend either the original or the altered version

---

[1]*Interactivity*, as defined by Biocca (1992), refers to "the number and forms of input and output, the level of responsiveness to ... user actions and states, and the range of interactive experiences (including applications) offered by the system" (p. 64).

to one or potentially thousands of others. Furthermore, users can also interact with others in real time and carry on a live conversation, an experience that is becoming increasingly richer with higher resolutions and sensory repleteness of various modalities of interactive technologies.

Temporal flexibility also means that users can choose to attend to the health information in a sequence and context determined by themselves. Once the information has been stored, users can control for themselves the pace at which it is presented. One of the extolled virtues of text, for example, is that it facilitates cognitive rehearsal, as it is self-paced (Singer, 1980). Self-pacing supposedly makes it possible to review material, scan forward and backward, and form mental images to enhance learning. Pacing, however, is a decision undertaken by the user or the viewer and is a property of the reading context; it is not inherent in the medium (see, for example, Wright et al, 1984, who define pace as the amount of information processed per unit time). Some media, such as television, provide no opportunities for the viewer for regulating the pace at which information is processed. Interactive technologies, on the other hand, leave pacing issues up to the user.

In the absence of opportunity for actual behavioral rehearsal, as is the case in typical television-viewing and computer-use situations, cognitive rehearsal of the desired behavior (the extent to which the media invite viewers to contemplate on what they have seen, heard, or read) can act as a powerful proxy (Maddux, 1993; Maibach & Flora, 1993). Many traditional media make it difficult to engage in cognitive processing, as the pace of presentation is usually controlled by the sender. Temporal flexibility of interactive media, however, enhances opportunities for cognitive rehearsal as users determine the pace, sequence, and the situational contexts under which messages will be processed.

The central theme here is one of heightened user control. Temporal flexibility increases the range of choices accorded the user. There is now a strong body of literature to suggest that providing individuals with choices enhances their sense of control (Langer, 1983). Perceived control in turn is associated with positive outcomes (Geer, Davidson, & Gatchel, 1970; Makowsky, Cook, Berger, & Powell, 1988), including higher levels of enjoyment (Cohen, 1981; Srull & Wyer, 1986), greater learning (Brigham, 1979; Perlmuter, Monty, & Kimble, 1971; Stine, Lachman, & Wingfield, 1993; Stipek & Weisz, 1981), and healthier behaviors (Bandura, 1989; Bradley, Gamsu, Moses, & Knight, 1987; Legg-England & Evans, 1992).

Although it is easy to get swept away by the virtues of user control, attention also needs to be paid to its potentially deleterious effects. Orton's

(1995) work suggests that perceived control interacts with goals associated with the interaction situation to determine enjoyment and interest. Whereas the psychological literature strongly endorses user control in learning and other educational endeavors (Bradley et al., 1987; Legg-England & Evans, 1992), work currently underway in narratives points toward the opposite: that providing control results in loss of interest and liking (Orton, 1995). In other words, the role of perceived control depends on whether the goal of the interaction is "I want to learn," or "Tell me a story." If the latter, users typically prefer abdication of control.

Thus, whereas education-oriented interactive technologies may benefit from providing user control, entertainment-oriented ones may not. From the perspective of using interactive technologies in health promotion, the need to distinguish between these two goals is not inconsequential. Relapse rates in smoking and alcohol use, for example, are high among individuals who, in the presence of alternatives, typically choose the more harmful ones (Malow, Pintard, Sutter, & Allain, 1988). This is not to suggest that we equate choices among alternatives in a behavior situation, such as smoking cessation, with those present in a user-technology interaction; it is to point to the necessity of understanding the conditions under which user control can be beneficial. A recent study finds, for example, that individuals perceive self-selected online news stories less credible or newsworthy than stories selected by other users, editors, or even computers (Sundar & Nass, 1996). In other words, users perceive themselves, relative to others, to be less able to select items of a technical nature. Health information, by implication, is best presented by a health professional.

Education and entertainment goals may also differ according to the levels of involvement associated with them. In a recent pilot test of breast cancer CD-ROM programs, Street and Manning (this volume) found that low, relative to high, involvement subjects preferred videotape (a modality with relatively lower levels of control) over a CD-ROM program (with higher levels of control) or brochures. Even though the experiment demonstrated a causal link between involvement as a cause and preference for control as an effect, we might also hypothesize the opposite chain of causality—that the presence of choices would enhance involvement. Involvement, in turn, is related to learning. Those with high, relative to low, levels of involvement tend to follow the central route of persuasion, paying attention to the arguments, as opposed to the peripheral cues, in a message (Petty & Cacioppo, 1981). As a result, learning rates are typically higher among high involvement subjects (Ray, 1973), who are also more likely to act in concordance with their knowledge and attitudes (Chaffee & Roser, 1986).

To the extent that user control can lead to greater involvement, it can result in higher rates of learning.

Finally, what users may choose to do with the host of options available in an interactive environment is another important consideration. Compared to light and nonchallenging information, difficult and challenging tasks require more expenditure of time and cognitive resources for processing. Faced with choices in such situations, users, especially those not committed to learning about a particular task, are likely to opt for the easy way out. Thus, although perceptions of control may be useful, and indeed desirable, in motivating people to participate in an endeavor, it may also be necessary to incorporate mechanisms that promote commitments once choices have already been made.

Message Tailoring Capabilities

Although, as with broadcast media, interactive technologies can also disseminate messages to multiple users from a single source, they have two comparative advantages. First, they make it possible to tailor messages much more precisely to a narrowly defined audience (those inflicted with a particular disease, for example, or caregivers). The same health information can be crafted to suit preferences and characteristics of targeted individuals. In other words, interactive technologies facilitate and improve on the sender's segmentation efforts.

The second and perhaps more important advantage of interactive technologies in message tailoring efforts is the primacy of users' preferences over those of the senders'. As mentioned earlier, the flood of information available through computers places a greater burden on individuals to act as their own gatekeepers. To combat this problem, many software programs and search engines have begun to install agents that select information according to criteria predetermined by users such as keywords, urgency of message, source of origin, and so forth. Thus, interactive technologies can promote effective message tailoring at both the sender and receiver ends of the communication process.

An implication of message tailoring capabilities of interactive technologies is the possibility for audience segmentation, defined as "the clustering of target subgroups according to a set of variables" (Cirksena & Flora, 1995, p. 212), which has been proposed to increase the reach and effectiveness of health promotion efforts (Slater & Flora, 1991). The more narrowly a message is designed to suit user characteristics, the more likely it is to promote learning (Greenfield, Kaplan, Ware, Yano, & Frank, 1988). Conversely, improvements in individuals' health decisions are likely to be minimal as long as messages are impersonal (Dede and Fontana, 1995).

Audience segmentation can be achieved in a number of ways. First, messages can be tailored to residents of particular geographical locations. Information about the availability of (health) services can be segmented by state, city, or even zip code, as is already done by many commercial enterprises.

Second, demographic characteristics can be used to segment audiences and tailor messages (Williams & Flora, 1995). Breast cancer screening, for example, may be of more interest to older women than to others. In this group, messages can be written for more narrowly defined subgroups, based, for example, on literacy levels, prior knowledge, or prior condition.

A third method of segmentation is through the psychobehavioral profile of the audience. The transtheoretical model (Prochaska & DiClemente, 1984; Prochaska et al., 1994) suggests that efforts to change addictive behaviors are more likely to succeed when the target audience is segmented according to its stage of change. Those classified as precontemplators for smoking cessation, for example, require a different strategy than those at the action stage, who have already undertaken cessation efforts (Velicer, Hughes, Fava, & Prochaska, 1995).

As pointed out earlier, the distinctive feature of interactive technologies is their ability to form unique combinations of various attributes for optimal benefit and effectiveness. Although television may contribute to the learning process by modeling the desired behavior, for example, it is a poor medium for providing feedback. Similarly, newspapers may enhance the processing of complex information (Kozma, 1991), but they are ineffective tools for promoting observational learning. Interactive technologies, on the other hand, can combine relevant modality features into an integrated unit. The extent to which individuals use interactive technologies for their health and well-being is yet to be known. What we can predict, however, is that utilization rates will likely be determined by the creativity with which health promotion efforts incorporate and take advantage of the various attributes of interactive technologies that promote learning and behavior change, many of which have been outlined in this chapter.

Using the THUS Framework

Having discussed important technology attributes, we next refer back to the THUS framework discussed at the beginning of this chapter and outline some future research directions for the study of interactive technologies in health promotion.

The first column in the framework comprises the attributes of technologies expected to impact health behaviors. Note that, because we list attributes and not the technology itself, this framework is not limited to interactive technologies, but applies to any entity possessing one or more of the attributes (Nass & Mason, 1990). Television, for example, would score high on multimodality, availability, and ease of use, but low on others. Similarly, a telephone would score high on interactivity and networkability but low on multimodality, and so on.

Theoretical development in health promotion has to move beyond the investigation of specific health domains and look for commonalities and differences across them (Maibach, Maxfield, Ladin, & Slater, 1996). The second column (see Table 2.1) lists some key health behavior attributes that comprise various health domains. Smoking cessation as a health domain, for example, is associated with long-term health benefits, and it would score high on addictiveness and maintenance of change; moderate on feasibility of self-help, skills requirement, and importance of social support; and low on knowledge requirement and role of heredity. Exercising, another health domain, has the attributes of low addictiveness, long-term health benefits, high feasibility for self-help, long-term maintenance of change, low knowledge and moderate skills requirements, moderate levels of importance of social support, and minimal roles of both heredity and confidentiality requirement (USDHHS, 1996).

The last column in Table 2.1 lists attributes of users likely to influence use and effects of interactive technologies for specific health behavior changes. Multimodality in column 1 will likely be more important for those with low levels of knowledge, relative to those whose prior knowledge about a particular health domain is already high. Hence, presentation of text-based information on blood sugar management (a health domain with high knowledge requirement), for example, will likely be more effective for those with high than low prior knowledge.

In this framework, vertical combinations define particular technologies (in the first column), health domains (second column), and users (third column); horizontal combinations define the THUS interactions. The benefit of the framework proposed here is its heuristic value. Each set of imaginary lines connecting the three columns comprises a potential research question about how different users react to or make use of different technologies for different health domains. We next illustrate the use of the THUS framework in two broad areas of research: adoption and effects.

Questions About Adoption.   In studying the potential and actual effects of technology, a fundamental set of questions asked by researchers concerns the diffusion of innovations (Rogers, 1962): How fast does the technological innovation diffuse in society, and which characteristics differentiate adopters from nonadopters? Although this question is primarily concerned with user attributes ("How do adopters differ from nonadopters?") and focuses on a particular technology (that is, the first column is held constant), answers to this question reside in both the first and third columns. In other words, a fuller understanding of the diffusion rates of a technology requires not only an understanding of user characteristics, but those of the technology as well.

Consider the costs associated with new technologies and concerns that have been raised about questions of access—that individuals without the financial wherewithal will be left further behind in health care (Sutton, 1991). Research investigating the differential utilization rates of new technologies across the economic strata in the population concerns attributes about technologies and individuals: What is the relationship between costs of a technology and the socioeconomic characteristics of its users?

Questions About Effects.   For some effects, attributes of individuals may be less consequential than those of the technology. The computers-as-social-actors paradigm (Nass, Moon, Fogg, & Reeves, 1995; Nass & Steuer, 1993) posits, for example, that human-computer interactions are fundamentally social, and that we unconsciously use social heuristics to guide our interactions with machines (Nass, Steuer, Henriksen, & Dryer, 1994). In this paradigm, user characteristics do not play a major role as the theoretical claim is not restricted to particular types of individuals or those with particular abilities. Instead, technology attributes are key, and the research question can be framed as: What attributes of technologies elicit social heuristics?

For other effects, all three sets of attributes are central issues. Research in this area can ask questions about one, two, or all three sets of attributes. A research question that asks, for example, "Are women likely to be influenced by online chat groups to learn about breast self-examination (BSE)?" could be investigated by generating a series of hypotheses about relationships among the various entries in the three columns: networkability (to connect to chat groups) and multimodality (to learn BSE skills) as attributes of technology; importance of social support, feasibility of self-help, and skills requirement as attributes of the health domain; and psychographics (e.g., attitudes and self-efficacy) as attributes of users. Similarly, the relative efficacy of two methods of treatment (CD-ROM versus videos, for example) can be assessed by specifying the relevant attributes of the two media (multimodality, temporal flexibility, interactivity, etc.), the nature of the desired outcome (short- versus long-term effects), and the relevant target audience characteristics (men versus women, those with high versus low self-efficacy, etc.).

Conclusion

Enactment of healthy behaviors, many of which involve the abdication of addictive lifestyle patterns (such as smoking) or the adoption of difficult tasks that have to be performed on a regular basis (exercising, for many people, for example), cannot be achieved by merely interacting with a computer. Determination, motivation, heightened sense of self-efficacy, realistic appraisals of extant barriers, willingness to learn and practice in the face of setbacks, and a supportive social environment are also required. In this chapter, we have not addressed many of these issues, but instead have argued that interactive technologies, if properly designed, can move individuals toward healthier lifestyles. To do so, we need a better understanding of the features of the technology, including its abilities and restrictions, characteristics of users, as well as the attributes of the health behavior. Only then can we expect to make full use of the technology in promoting health.

REFERENCES

Allen, A. (1994). Teleradiology 1994. *Telemedicine Today, 2*(1), 21–22.
Anderson, D. R., Alwitt, L. F., Lorch, E. P., & Levin, S. R. (1979). Watching children watch television. In G. A. Hale & M. Lewis (Eds.), *Attention and cognitive development* (pp. 331–361). New York: Plenum.
Anderson, D. R. & Levin, S. R. (1976). Young children's attention to *Sesame Street. Child Development, 47,* 806–811.
Bandura, A. (1977). *Social learning theory.* Engelwood Cliffs, NJ: Prentice-Hall.
Bandura, A. (1986). *Social foundations of thought and action.* Engelwood Cliffs, NJ: Prentice-Hall.
Bandura, A. (1989). Regulation of cognitive processes through perceived self-efficacy. *Developmental Psychology, 25,* 729–735.
Bates, J. (1992). Virtual reality, art, and entertainment. *Presence, 1,* 133–138.
Biocca, F. (1992). Communication within virtual reality: Creating a space for research. *Journal of Communication, 42,* 5–22.
Blumler, J. G., & Katz, E. (Eds.). (1974). *The uses of mass communications: Current perspectives on uses and gratifications research.* Newbury Park, CA: Sage.

Boberg, E. W., Gustafson, D. H., Hawkins, R. P., Chan, C., Bricker, E., Pingree, S., & Berhe, H. (1995). Development, acceptance, and use patterns of a computer-based education and social support system for people living with AIDS/HIV infection. *Computers in Human Behavior, 11*, 289–311.

Bradley, C., Gamsu, D. S. Moses, J. L., & Knight, G. (1987). The use of diabetes-specific perceived control and health belief measures to predict treatment choice and efficacy in a feasibility study of continuous subcutaneous insulin infusion pumps. *Psychology & Health, 1*, 133–146.

Brennan, P. F., Moore, S. M., & Smyth, K. S. (1995). The effects of a special computer network on caregivers of persons with Alzheimer's Disease. *Nursing Research, 44*, 166–172.

Brigham, T. (1979). Some effects of choice on academic performance. In L. C. Perlmuter & R. A. Monty (Eds.), *Choice and perceived control* (pp. 131–141). Hillsdale, NJ: Lawrence Erlbaum Associates.

Brookfield, S. (1986). Media power and the development of media literacy: An adult educational interpretation. *Harvard Educational Review, 56*, 151–170.

Chaffee, S. H., & Roser, C. (1986). Involvement and the consistency of knowledge, attitudes, and behavior. *Communication Research, 13*, 373–399.

Chamberlain, M. A. (1994). New technologies in health communication: Progress or panacea? *American Behavioral Scientist, 38*(2) 271–284.

Cirksena, M. K., & Flora, J. A. (1995). Audience segmentation in worksite health promotion: A procedure using social marketing concepts. *Health Education Research, 10*, 211–224.

Cohen, C. E. (1981). Goals and schemata in person perception: Making sense from the stream of behavior. In N. Cantor & J. F. Kihlstrom (Eds.), *Personality, cognition, and social interaction* (pp. 45–68). Hillsdale, NJ: Lawrence Erlbaum Associates.

Crawford, C. M., & HIAWG Work Group on Managed Care. (1996). *Managed care and the NII: A public/private perspective.* Information Infrastructure Task Force, Committee on Applications and Technology, Health Information Applications Work Group. Final Draft.

Dede, C., & Fontana, L. (1995). Transforming health education via new media. In L. Harris (Ed.), *Health and the new media* (pp. 163–183). Mahwah, NJ: Lawrence Erlbaum Associates.

America Offline. (1996, June 29). *The Economist*, p. 64.

Elmer-Dewitt, P., Jackson, D. S., & Ratan, S. (1994, July 25). Battle for the soul of the Internet. *Time, 144*, pp. 50–56.

Famighetti, R. (Ed.). (1996). Computers: The internet (p. 167). *The world almanac and book of facts.* Mahwah, NJ: World Almanac Books.

Ferguson, T. (1996). *Health Online: How to go to find health information, support groups, and self-help communities in Cyberspace.* Addison-Wesley.

Geer, J. H., Davidson, G. C., & Gatchel, R. J. (1970). Reduction of stress in humans through nonveridical perceived control of aversive stimulation. *Journal of Personality and Social Psychology, 16*, 731–738.

Greenfield, S., Kaplan, S. H., Ware, J. E., Yano, E. M., & Frank, H. J. L. (1988). Patients' participation in medical care: Effects on blood sugar control and quality of life in diabetes. *Journal of General Internal Medicine, 3*, 448–457.

Gustafson, D., Wise, M., McTavish, F., Taylor, J. O., Smalley, R., Wolberg, W., Stewart, J. (1993). Development and pivotal evaluation of a computer-based support system for women with breast cancer. *Journal of Psychosocial Oncology, 11*, 69–93.

Kozma, R. M. (1991). Learning with media. *Review of Educational Research, 61*, 179–211.

Langer, E. J. (1983). *The psychology of control.* Beverly Hills, CA: Sage.

Legg-England, S., & Evans, J. (1992). Patients' choices and perceptions after an invitation to participate in treatment decisions. *Social Science & Medicine, 34,* 1217–1225.

Maddux, J. E. (1993). Social cognitive models of health and exercise behavior: An introduction and review of conceptual issues. *Journal of Applied Sport Psychology, 5,* 175–183.

Maibach, E. & Flora, J. A. (1993). Symbolic modeling and cognitive rehearsal: Using video to promote AIDS prevention self-efficacy. *Communication Research, 20,* 517–545.

Maibach, E. W., Maxfield, A., Ladin, K., & Slater, M. (1996). Translating health psychology into effective health communication: The American healthstyles audience segmentation project. *Journal of Health Psychology, 1,* 261–277.

Makowsky, P. P., Cook, A. S., Berger, P. S., & Powell, J. (1988). Women's perceived stress and well-being following voluntary and involuntary relocation. *Lifestyles, 9,* 111–122.

Malow, R. M., Pintard, P. F., Sutter, P. B., & Allain, A. N. (1988). Psychopathology subtypes: Drug use motives and patterns. *Psychology of Addictive Behaviors, 2,* 1–13.

McGuire, W. J. (1989). Theoretical foundations of campaigns. In R. E. Rice, & C. Atkin (Eds.), *Public communication campaigns* (pp. 43–65), Newbury Park, CA: Sage.

Nass, C. I., & Mason, L. (1990). On the study of technology and task: A variable-based approach. In J. Jult & C. Steinfeld (Eds.), *Organizations and communication technology* (pp. 46–67). Newbury Park, CA: Sage.

Nass, C. I., Moon, Y., Fogg, B. J., & Reeves, B. (1995). Can computer personalities be human personalities? *International Journal of Human-Computer Studies, 43,* 223–239.

Nass, C. I., & Steuer, J. (1993). Voices, boxes, and sources of messages: Computers and social actors. *Human Communication Research, 19,* 504–527.

Nass, C. I., Steuer, J., Henriksen, L., & Dryer, D. C. (1994). Machines, social attributions, and ethopoeia: Performance assessments of computers subsequent to "self-" and "other-" evaluations. *International Journal of Human-Computer Studies, 40,* 543–559.

Negroponte, N. (1995). *Being digital.* New York: Knopf.

Newhagen, J. E., Cordes, J. W., & Levy, M. R. (1995). Nightly@nbc.com: Audience scope and the perception of interactivity in viewer mail on the Internet. *Journal of* Communication, *45,* 164–175.

Newhagen, J. E. (1994). Self efficacy and call-in political television show use in 1992. *Communication Research, 21,* 366–379.

Office of Rural Health Policy. (1994). *Reaching Rural.* Report of the U.S. Department of Health and Human Services, Health Resources and Services Administration, Washington, DC.

O'Leary, S., Mann, C., & Perkash, I. (1991). Access to computers for older adults: Problems and solutions. *The American Journal of Occupational Therapy, 45,* 636–642.

Orton, P. (1995). *Effects of perceived choice and narrative elements of interest in liking of story.* Unpublished doctoral dissertation, Stanford University, CA.

Perlmuter, L. C., Monty, R. A., & Kimble, G. A. (1971). Effect of choice on paired-associate learning. *Journal of Experimental Psychology, 91,* 47–53.

Perry, C. L., Baranowski, T., & Parcel, G. S. (1990). How individuals, environments, and health behavior interact: Social learning theory. In K. Glanz, F. M. Lewis, & B. K. Rimer (Eds.), *Health behavior and health education: Theory, research, and practice* (pp. 161–186). San Francisco, CA: Jossey-Bass.

Petty, R., & Cacioppo, J. T. (1981). *Attitudes and persuasion: Classic and contemporary approaches.* Dubuque, IA: Brown.

Pingree, S., Hawkins, R. P., Gustafson, D. H., Boberg, E. W., Bricker, E., Wise, M., & Tillotson, T. (1993). Will HIV-positive people use an interactive computer system for information and support? A study of CHESS in two communities. *Proceedings of the seventeenth Annual Symposium on Computer Applications in Medical Care, 17,* 22–26.

Prochaska, J. O., & DiClemente, C. C. (1984). *The transtheoretical approach: Crossing traditional boundaries of change.* Homewood, IL: Dow Jones-Irwin.

Prochaska, J. O. Velicer, W. F., Rossi, J. S., Goldstein, M. G., Marcus, B. H., Rakowski, W., Fiore, C., Harlow, L. L., Redding, C. A., Rosenbloom, D., & Rossi, S. R. (1994). Stages of change and decisional balance for 12 problem behaviors. *Health Psychology, 13,* 39–46.

Rafaeli, S. (1988). Interactivity: From new media to communication. In R. P. Hawkins, J. M. Wiemann, & S. Pingree (Eds.), *Advancing communication science: Merging mass and interpersonal processes* (pp. 110–134). Newbury Park, CA: Sage.

Ray, M. L. (1973). Marketing communication and the hierarchy-of-effects. In P. Clarke (Ed.), *New models for communication research* (pp. 147–176). Beverly Hills, CA: Sage.

Reeves, B. (1991). *"Being there": Television as symbolic versus natural experience.* Unpublished manuscript, Institute of Communication Research, Stanford University, Stanford, CA.

Rogers, E. M. (1986). *Communication technology: The new media in society.* New York: Free Press.

Rogers, E. M. (1962). *Diffusion of innovations.* New York: The Free Press.

Ross, J. A. (1995). Effects of feedback on student behavior in cooperative learning groups in a grade 7 math class. *Elementary School Journal, 96,* 125–143.

Ross, S. M., & Rakow, E. A. (1981). Learner control versus program control as adaptive strategies for selection of instructional support on math rules. *Journal of Educational Psychology, 73,* 745–753.

Rusted, R., & Coltheart, V. (1979). The effect of pictures on the retention of novel words and prose passages. *Journal of Experimental Child Psychology, 28,* 516–524.

Shannon, C. E., & Weaver, W. (1949). *The mathematical theory of communication* (pp. 95–117). Urbana: University of Illinois Press.

Sheridan, T. B. (1992). Musings on telepresence and virtual presence. *Presence: Teleoperators and virtual environments, 1,* 120–126.

Singer, J. L. (1980). The power and limitations of television: A cognitive–affective analysis. In P. H. Tannenbaum (Ed.), *The entertainment functions of television* (pp. 31–65). Hillsdale, NJ: Lawrence Erlbaum Associates.

Singer, J. L., & Singer, D. G. (1979, March). Come back, Mister Rogers, come back. *Psychology Today,* pp. 56, 59–60.

Slater, M., & Flora, J. A. (1991). Health lifestyles: Audience segmentation for public health interventions. *Health Education Quarterly, 18,* 221–234.

Srull, T. K., & Wyer, R. S., Jr. (1986). The role of chronic and temporary goals in social information processing. In R. M. Sorrentino & E. T. Higgins (Eds.), *Handbook of motivation and cognition* (pp. 503–549). New York: Guilford.

Steuer, J. (1992). Defining virtual reality: Dimensions determining telepresence. *Journal of Communication, 42,* 73–93.

Stine, E. A., Lachman, M. E., & Wingfield, A. (1993). The roles of perceived and actual control in memory for spoken language. *Educational Gerontology, 19,* 331–349.

Stipek, D. J., & Weisz, J. R. (1981). Perceived personal control and academic achievement. *Review of Educational Research, 51,* 101–137.

Stone, D., & Glock, M. (1981). How do young adults read directions with and without pictures? *Journal of Educational Psychology, 73,* 419–426.

Sundar, S. S., & Nass, C. (1996, May). *Source effects in users' perception of online news.* Paper presented at the 46th Annual Conference of the International Communication Association's annual meeting, Chicago, IL.

Sutton, R. E. (1991). Equity and computers in the schools: A decade of research. *Review of Educational Research, 61,* 475–503.

Untangling the Web: Minorities strive to catch up in cyberspace (July, 1995). *Minority Markets ALERT, 58*(4), 4.

U.S. Department of Health and Human Services. (1996). *Physical activity and health: A report of the Surgeon General.* Atlanta, GA: U.S. Department of Health and Human Services, Centers for Disease Control and Prevention, National Center for Chronic Disease Prevention and Health Promotion.

Velicer, W. F., Hughes, S. L., Fava, J. L., & Prochaska, J. O. (1995). An empirical typology of subjects within stage of change. *Addictive Behaviors, 20,* 299–320.

Wallace, S.T., & Alden, L. E. (1995). Social anxiety and standard setting following social success or failure. *Cognitive Therapy & Research, 19,* 613–631.

Walther, J. B., Anderson, J. F., & Park, D. W. (1994). Interpersonal effects in computer-mediated interaction: A meta-analysis of social and antisocial communication. *Communication Research, 21,* 460–487.

Watts, K., Baddeley, A., & Williams, M. (1982). Automated tailored testing: The Mill Vocabulary tests: A comparison with manual administration. *International Journal of Man Machine Studies, 17,* 241–246.

Williams, J. E., & Flora, J. A. (1995). Health behavior segmentation and campaign planning to reduce cardiovascular disease risk among Hispanics. *Health Education Quarterly, 22,* 36–48.

Wright, J. C., Huston, A. C., Ross, R. P., Calvert, S. L., Tolandelli, D., Weeks, L. A., Raeissi, P., & Potts, R. (1984). Pace and continuity of television programs: Effects on children's attention and comprehension. *Developmental Psychology, 20,* 653–666.

3

Using Theories in Planning Interactive Computer Programs

Celette Sugg Skinner
Washington University School of Medicine

Matthew W. Kreuter
Saint Louis University School of Public Health

Interactive media technology provides a unique opportunity for tailoring health behavior change messages for individuals in a way that mass-produced brochures or videos cannot. However, the potential of interactive computer programs to serve this function has not yet been realized. One explanation for the small number of interactive programs with demonstrated efficacy is that few individuals possess the expertise necessary to develop them singlehandedly. To effectively use interactive technology in promoting health-related behavior change, program developers must understand:

1. the nature of the target behavior or health problem and its determinants among members of a specified population;
2. the processes of behavior change, including knowledge and application of a range of theories that address the determinants of the target behavior or health problem; *and*
3. interactive computer technologies and communication strategies, and how to combine them to create an effective health promotion program for the population of interest.

Because lack of expertise in any one of these areas will probably result in less-than-optimal programs, it is important for development teams to include not just health educators, but also communication specialists, behavioral scientists, education specialists, computer programmers, graphic artists, and others who bring unique and essential skills to this process. It is equally important, we think, that all team members recognize the value and inter-dependence of each distinct part of the development process.

The purpose of this chapter is twofold: First, we describe how behavioral science theories can be used to enhance the effectiveness of health promotion campaigns in reaching their objectives. We do this within the context of a program-planning framework for identifying determinants of health-related behaviors. Five behavior-change theories are outlined along with discussion of how and when each might be applied. Second, we overview how the various theoretical frameworks can be used in interactive technology to tailor health messages to the particular needs, interests, and life situations of the targeted users.

TYPES AND DETERMINANTS
OF TARGET BEHAVIORS

Several different types of health-related behaviors may be targeted for change. Some of these are delineated in Table 3.1. Skills and psychological factors affecting one behavior may differ from those affecting another.

Before beginning a health-promotion intervention, one should carefully study the type and the determinants of the particular behavior the program is seeking to change. This is sometimes referred to as conducting a behavioral diagnosis. The behavioral diagnosis should suggest a relevant theoretical framework for conceptualizing the behavior-change process. This

TABLE 3.1
Types of Health-Related Behaviors

Behavior Type	Example	Interventions
Asymptomatic screening	pap smears, colorectal screening	Ask people who are not sick and have no apparent problems to interact with the health care system for early disease indentification.
Lifestyle modifications	diet, exercise	Seek to change habitual behaviors involved in everyday life. Members of target population may be people who have no apparent health problems.
Cessation of addictive behaviors	alcohol, tobacco use	Encourage maintained abstinence. Recommended behavior change is to stop a frequent behavior rather than adopt a new one.
Medical regimen compliance	pill taking, glucose monitoring	Encourage adherence to regimens prescribed for specific diseases or medical conditions.
Precaution adoption	radon testing, smoke detector installation	Encourage one-time or infrequent activities to protect against a threat that may be viewed as unlikely.

chapter outlines a sample of five theoretical frameworks for behavior change, and links them with an over-arching planning model for behavior-change interventions.

UNDERSTANDING PROCESSES
OF BEHAVIOR CHANGE

This section discusses the health belief model, efficacy expectations (from social cognitive theory), attribution theory, the theory of reasoned action, and the transtheoretical model. These prominent theories of health-related behavior change were chosen because studies have demonstrated not only their value in predicting health-related behavior change, but also the effectiveness of interventions based upon them. For each of these theories, we will provide a description of the theory and its constructs, a discussion of how the theory relates to the PRECEDE/PROCEED planning model, and applications for interactive computer interventions.

The Health Belief Model

A Description.   The health belief model (HBM; Fig. 3.1) is one of the most widely used models for explaining and predicting individuals' health behavior change. The model grew out of investigations in the 1950s, when U.S. Public Health Service officials were struggling to understand why people were not using tuberculosis screening units conveniently located in their neighborhoods. Social psychologists engaged to study the problem identified variables that interact to influence a person's health behaviors. These variables were delineated in the HBM (Hochbaum, 1958; Janz & Becker 1984; Kirscht & Rosenstock 1979).

The HBM is a value-expectancy model in which "behavior is seen as a function of the subjective value of an outcome and of the subjective probability or expectation that a particular action will achieve that outcome" (Rosenstock, 1991, p. 40). The model does not suggest that behavior change is a function of knowledge or understanding alone. Rather, it recognizes that, until one expects some value in making a behavior change, there will be no reason to even consider the change. The HBM postulates that health behaviors are influenced by perceptions of severity, susceptibility, benefits, and barriers associated with a health action and that other factors and cues to action also play a role (Kirscht & Rosenstock, 1979, Rosenstock, 1991).

The HBM postulates that people are more likely to engage in a health action if they think they are at risk for a condition they consider to be severe,

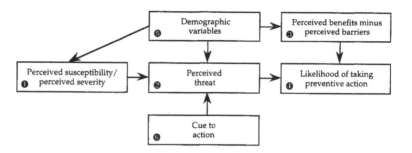

FIG. 3.1.   The Health Belief Model.

the health action can protect them against the threat, and the barriers to doing so do not outweigh the potential benefits.

- *Perceived severity* (❶) refers to beliefs about the nature of a health threat. Both social and medical consequences are involved here. In short, this construct concerns "how bad" a person thinks it would be to develop a certain health-related condition.
- *Perceived susceptibility* (❶) concerns the subjective perception of the risk of contracting the health condition. A person's perceived likelihood of experiencing the health threat may be very different from actual, epidemiologic risk.
- *Perceived threat* (❷) of the health-related condition is determined by perceived susceptibility and severity. When a person believes he or she is at higher risk (susceptibility) for a serious problem (severity), then perceived threat is high. In the HBM, perceived threat is considered a necessary precursor to behavior change
- *Perceived benefits* (❸) have to do with a specific protective health behavior being considered. They are beliefs that the specific action will have benefits in protecting against the particular threat. In assessing whether there is a benefit associated with an action, the basic question is, "Will this action help me avoid this health threat?"
- *Perceived barriers* (❸) associated with taking an action are potential negative aspects. It is thought that, when a person considers whether to act, barriers are nonconsciously weighed against the potential benefits. Expense, danger, pain, inconvenience, and disapproval from relevant others are examples of potential barriers discouraging action.
- *Other variables* (structural, sociopsychological and demographic; ❺) may affect people's perceptions of the variables outlined here and, thus, may indirectly influence health behavior (Rosenstock, 1991).

- *Cues* (◉) can affect a person's readiness to take action. Near collision can lead to seat belt use; a physical symptom can trigger a delayed physician visit. Presumably, a health promotion campaign will act as a cue to health action.

Applications for Behavior Change.   Initially applied to behaviors relating to screening and compliance (Strecher & Rosenstock, 1996), the HBM has been the basis for a plethora of health behavior and intervention research studies. There is neither an accepted method of determining the relative weight of each HBM variable nor a specified method for predicting behavioral outcomes given various interactions of variable values (Wallston & Wallston, 1984). That was not the developers' original intent. Rather, the model delineates variables that may influence individuals' behaviors. In 1984, Janz and Becker reviewed studies using the HBM over the previous decade. Perceived barriers was the most powerful single predictor of health actions across all studies. Perceived risk (or susceptibility) was a stronger predictor of participation in preventive health behaviors, whereas perceived benefits was a stronger predictor of health actions recommended for people who already had a medical condition, such as compliance with medicine taking (Janz & Becker, 1984).

HBM variables are important for people's decisions about whether to even consider or try to make a behavior change. Perceptions specified in the HBM may be necessary but not sufficient for a decision to change behavior. For instance, although perceived risk and belief in the efficacy or benefits of screening do not necessarily cause people to participate in cancer screening, there will be little reason to even consider screening in the absence of these perceptions. Whether the decision actually gets carried out and sustained can depend on the difficulty and extent of the change. The perceived relative weight of benefits and barriers is important at this point. Summary findings from a 1994 review of 13 mammography utilization studies based on the HBM are consistent with this prediction; across the studies, perceived risk was associated with mammography use but was not as strongly predictive as were perceived barriers to having a mammogram (Curry & Emmonds, 1994).

Efficacy Theory (From Social Cognitive Theory)

A Description.   If persons do not expect that a health action will yield a beneficial result, they have little reason to act. Similarly, if they do not expect to be able to take the action, they will have little reason to attempt it. Social cognitive theory (Baranowski, Perry, & Parcel, 1996), (formerly titled social learning theory; Bandura, 1977a), suggests that behavior change and maintenance are a function of these two types of expectations—outcome expectations (whether a certain behavior will lead to a certain outcome) and self-efficacy expectations (whether one is capable of successfully engaging in the behavior; see Fig. 3.2). As is true of HBM constructs, it is perceived efficacy—not actual capability—that is important. Outcome expectations are analogous to perceived benefits in the Health Belief Model and lack of perceived self-efficacy may be seen as a barrier to health behavior change. Self-efficacy expectations have been specified as an additional component in the HBM (Rosenstock, Strecher, & Becker, 1988); see Fig. 3.2.

- The likelihood that a person will execute a particular behavior is largely a function of two factors: self-efficacy and outcome expectations.
- *Self-efficacy* expectations refer to beliefs a person has about his or her ability to perform a behavior (❶). For example, does the recovering alcoholic believe he is capable of ordering a soft drink when his friends are all drinking beer?
- *Outcome expectations* are a person's beliefs about whether a behavior will lead to some specific outcome (❷). For example, does the person believe that by ordering a soft drink, he will be less likely to have a drink of beer?
- A person is most likely to perform some behavior when self-efficacy is high and outcome expectations are positive.
- Persons with high self-efficacy for using coping responses in high-risk situations are more likely to do so and they are more likely to maintain the behavior change (less likely to relapse).

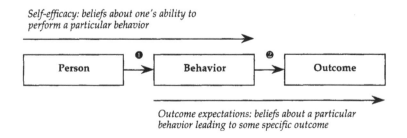

FIG. 3.2.   Self-efficacy (from social cognitive theory).

- Persons who successfully use a coping response in a high-risk situation will have even greater self-efficacy to do the same in the future.
- Persons with positive outcome expectations (e.g., perceive desirable consequences of not having a drink) are more likely to maintain positive behavior change.

Perceived self-efficacy affects all kinds of behaviors but may vary from one type of behavior to the next (i.e., self-efficacy is not a global personality trait). Thus, it is important to assess perceived efficacy for each specific behavior in question. Perceived efficacy may affect not only expectations for success, but also choices about where to attempt the task and how long or hard one works on the task (Strecher, DeVellis, Becker, & Rosenstock, 1986).

According to Bandura (1977a), there are four major sources for self-efficacy expectations—performance accomplishments, vicarious experience, verbal persuasion, and physiological state. In performance accomplishments, individuals learn through personal experience. Performance accomplishments are thought to be the strongest source for enhancing efficacy, perhaps because skills are being strengthened while expectations of further success are enhanced. In vicarious experience, people learn through observing others. Witnessing another person accomplishing a task (or "modeling" the desired behavior) may enhance one's own perceived efficacy, especially if the person who succeeds is perceived as being similar to the observer. Verbal persuasion is the traditional method of advancing information or arguments meant to persuade a person that he or she is capable of changing behavior. Finally, physiological cues can affect perceived efficacy (e.g., people are less likely to perceive self-efficacy when they become nervous, breathless, or begin to perspire when attempting a task).

In sum, an intervention based on efficacy theory should emphasize both self-efficacy and outcome-efficacy. Self-efficacy expectations are related to success of behavior-change attempts but, in initial stages of consideration, attempts will not even be considered if outcome expectations are not strong.

Applications for Behavior Change.   In general, high self-efficacy has been shown to predict success for change in a wide range of health-related behaviors (Strecher et al., 1986). Persons with higher self-efficacy are more likely to initiate new behaviors, maintain behavior they have already changed, put forth greater effort toward changing, and persist longer in their efforts (Bandura, 1977a, 1982b). For instance, studies of recovering alcoholics have found that those who had higher ratings of self-efficacy for their ability to resist temptations to drink in high-risk situations had lower rates of relapse (Annis & David, 1988; Chaney, O'Leary, & Marlatt, 1978).

Relapse prevention intervention strategies suggested by efficacy theory include teaching general and specific coping skills to deal with high-risk situations, building self-efficacy by rehearsing coping responses, starting with simple coping responses, focusing on positive aspects of an incomplete coping response, providing reinforcement, and changing outcome expectations about the effects of the behavior.

Attribution Theory

A Description.   Attribution theory (Fig. 3.3) is concerned with the explanations people give to health-related events and issues in their lives (Weiner, 1986). Central to the theory is the assumption that people ask "Why?"—why things happen to them, why they develop medical conditions, why they feel the way they do, or why they are at risk for a particular disease. Different answers to these "Why?" questions can affect coping with medical conditions and health behaviors. Attributions for past events can also affect expectations for the future (Strecher et al., 1986).

Types of attributions can be classified along four broad dimensions (or loci): causation, controllability, stability, and globality (Weiner, 1986).

- *Causation* can be internal or external. Internal causes have to do with the person who experienced the condition, event, or circumstance. External causes are outside the person (i.e., chance, luck, or something about the situation).
- *Controllability* simply denotes the extent to which the cause is believed to be controllable, whether by self or another.
- *Stability* concerns whether the cause is mutable. Unstable causes are changeable. Stable causes are unchangeable (by anyone or anything).

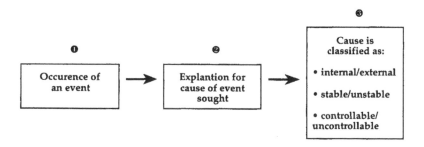

FIG. 3.3.   Attribution theory.

- *Globality* is the fourth dimension of causal attribution and concerns whether the cause affects only a limited set of outcomes (specific) or many outcomes (global). When intervening for one specific health behavior, this dimension may not be as relevant as the others. For this reason, our discussion will focus on the first three dimensions.

More than one cause can be attributed to a single outcome. Each of the mutually exclusive domains apply to each attributed cause. For instance, cause attributed to luck or chance would be external, uncontrollable, and changeable. Cause attributed to lack of effort would be internal, controllable, and changeable. Whether the attributed cause was perceived as affecting everything or only some things would determine whether it was specific or global.

Attribution type may be related to self-efficacy. For instance, a success will boost self-efficacy expectations only if it is attributed to one's own ability rather than to external or uncontrollable factors (Bandura, 1977b). Alternatively, in the case of failure, a person with high self-efficacy expectations may be more likely to attribute the cause to an external, unstable, uncontrollable cause such as chance in order to maintain the success expectation (Bandura, 1982a, 1982b).

- When an unexpected or negative event occurs people tend to search for an explanation (i.e., attribution) of the cause of the event.
- The explanations people generate can be classified along these dimensions: internal–external, stable–unstable, and controllable–uncontrollable
- An *internal* attribution identifies the cause of the negative event as something within the person. A recovering alcoholic who attributes a lapse to her lack of willpower instead of to something about the high-risk situation in which she lapsed is making an internal attribution.
- A *stable* attribution suggests that the cause of the event will always be present—it is a constant. If the recovering alcoholic believes that lack of willpower is just part of her personality, this attribution would be considered stable.
- An *uncontrollable* attribution suggests that the person making the attribution does not believe that the cause of the event is changeable.

Applications for Behavior Change. Attributions can affect mood, coping, and behavior. Attribution theory is seldom the single guiding framework for health behavior interventions but attempts to redirect attributions can be important components of an intervention strategy. Attributions can effect behavior change in the following way. Attributions for perceived risk to a health threat can effect whether an individual sees benefit (or efficacy) in a health action. People who are convinced the cause of their risk is stable will see no reason to participate in risk-reduction activities; neither will people who believe risk is due to unchangeable and/or uncontrollable causes. Intervention programs should focus users on risk causes that are changeable and controllable, such as heart disease risk modification by smoking cessation. In cases where disease risk is unchangeable (as in the case of genetic predisposition to breast cancer), an intervention should focus on reduction of mortality risk (through early detection and treatment facilitated through mammography screening). If the goal of the program is to facilitate a certain health behavior in the user, it might seek to foster a belief that the desired outcome can be brought about by something within the person's control (i.e., the cause of the outcome is internal, unstable, and controllable).

For lifestyle changes such as weight loss, interventions may focus on attributions for previous failures. The meaning given to previous failed behavior change attempts can strongly affect expectations for success in future attempts, and these expectations can, in turn, affect actual chance for success (Hospers, Kok, & Strecher, 1990). If previous failures are attributed to unstable (changeable) causes and/or previous successes are attributed to stable causes (not changed since last time nor likely to change in the future), there will be more expectation for success. Lower success expectations will result from previous successes being attributed to unstable causes and/or previous failures attributed to stable causes. Not surprisingly, the more previous failures individuals have experienced, the greater the likelihood they will attribute the failures to stable causes (Hospers, Kok, & Strecher, 1990).

For changes associated with addictive behaviors, internal, stable, and uncontrollable attributions for the causes of a lapse in abstinence increase the risk that a lapse in behavior maintenance will become a full-blown relapse. Using reattributional training, one would encourage the recovering alcoholic not to self-blame for a lapse, but to consider the high-risk situation as the primary cause. By identifying the situation, coping responses can be generated for future use.

Theory of Reasoned Action

A Description.   The theory of reasoned action (TRA; Fig. 3.4; Ajzen & Fishbein, 1980) assumes that humans reasonably process information when making behavioral decisions, and that most behaviors of social relevance are under individuals' volitional control. According to this theory, behaviors (**❻**) are mainly determined by intentions (**❺**). Whether people intend to engage in a behavior is a mathematical function of their own attitude toward the behavior (**❷**); the outcome they expect will result from the behavior multiplied by their evaluation of the desirability of that outcome; (**❶**) plus subjective norms (**❸**); what they think relevant others think about their engaging or not engaging in the behavior multiplied by their motivation to comply with those persons (**❹**). The relative importance of subjective norms and one's own attitudes may vary from individual to individual.

In simplistic form, the theory can be summed up as follows: Barring unforeseen events, people are likely to actually do what they intend to do; whether they intend to do something is determined by their own attitude about doing it and what they think others think about their doing it. Recognition that intentions are based on an entire set of attitudes and subjective norms helps explain why changing one or two beliefs will not necessarily result in behavior change. In addition, because a number of salient others may influence a person's intentions, belief that one other person (e.g., parent) approves of a behavior will not necessarily provide enough impetus for action—especially if others with whom the individual feels a stronger motivation to comply (e.g., friends) do not endorse the behavior.

The TRA introduces some important constructs that are consistent with, but are not specifically delineated, in the other models. In particular, the consideration of others' opinions, and the relative weight of those opinions on behavioral intentions based on motivation to comply, are important considerations. People who believe that salient others think they should, for instance, stop smoking, will feel some social pressure to do so. This pressure is likely to have more effect if they are motivated to comply with those who

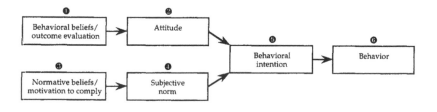

FIG. 3.4. Theory of reasoned action.

think they should stop smoking. In addition, there is a recognition that attitudes toward a behavior include not only whether a person perceives benefits associated with the behavior, but also the subjective perceived value of the benefit. Two people might both believe that smoking cessation will prolong life; one person might value a few extra years much more than the other.

Applications for Behavior Change.   The TRA, based as it is on rational behavioral choices, specifies that efforts to change behavior should be directed at changing people's beliefs. To use the model for intervention, one must first assess and then target the major beliefs on which attitudes are based and the perceived subjective norms influencing behavior. The TRA thus expands the HBM and efficacy theory's concepts of perceived benefits, barriers, severity, and efficacy by considering not only attitudes and perceptions, but also their relative weight determined by attitudes about the expected outcome and motivation to comply with opinions of salient others.

To change the attitudinal component:
Example—Changing attitudes about smoking to be less favorable

1. Add a new salient negative belief.
   (Cigarettes make you smell bad.)
2. Increase the unfavorability of an existing belief.
   (The cost of cigarettes is going up again.)
3. Increase the belief strength of an existing negative belief.
   (With today's smog and pollution, now more than ever before, it is especially hazardous for you to smoke.)
4. Decrease the favorability of an existing positive belief.
   (You don't look cool at all.)
5. Decrease the belief strength of an existing positive belief.
   (Smoking actually causes some people to gain weight.)
6. Focus only on certain beliefs at the exclusion of others.

To change the normative component :
Example—Changing attitudes about smoking to be less favorable

1. Add a new salient referent.
   (What would the woman at the health club think?)
2. Change the normative belief attributed to a reference.
   (You're wrong. John thinks smoking is stupid and that you should quit.)

3. Change the motivation to comply with a current referent.
   (Why do you care what he thinks, he doesn't even like you.)

To change the relative weights of attitude and normative beliefs:
Example—Changing attitudes about smoking to be less favorable

1. Emphasize the component that is most favorable to the desired intention.
   (Of course your friends don't want you to quit smoking; they all smoke, too. But it's your decision, not theirs.)

Unfortunately, more studies have used the TRA to strictly test the predictive probability of its mathematical formula rather than to evaluate the effect of tailoring interventions to recipients' specific attitudes and subjective norms (Curry & Emmonds, 1994). For instance, the model has been used to study intentions to engage in breast cancer control behaviors including breast self-examination, in which attitudes predicted about a third of variation in intentions whereas social norms were not predictive (Lierman, Young, Kasprzyk, & Benoliel, 1990); seeking medical attention for a symptom potentially indicative of breast cancer, for which attitudes and subject norms predicted more than half of the variance in intentions, but of the two, attitudes were much stronger determinants (Timko, 1987); and screening mammography, in which attitudes and subjective norms predicted intention and behavior (Montano & Taplin 1991).

Transtheoretical Model

A Description.   Traditional study of health behaviors dichotomizes outcomes according to whether the person did or did not engage in a certain behavior (i.e., did/did not have a pap smear, begin wearing seatbelts every time in a car, maintain smoking abstinence for a given period, etc.). However, preventive health behaviors may be more appropriately conceptualized as processes of adoption, analogous to diffusion of innovations (Rogers, 1962). Stage models employ this process concept by considering not only whether individuals have or have not performed a preventive health action, but also the degree to which they have considered the action and whether they have repeated it when appropriate. This approach allows for identification of variables predicting contemplation and relapse as well as action.

The transtheoretical model (Fig. 3.5; Prochaska & DiClemente, 1988), a widely used stage model, was originally developed as a conceptual tool for considering the process of smoking cessation (Prochaska & DiClemente, 1983). The model also was been found to be useful for a variety of health behaviors as diverse as alcohol abstinence (DiClemente & Hughes, 1990), sun screen use (Rossi, 1989), dietary change (Glanz et al., 1994) and contraceptive use (Grimley, Riley, Bellis, & Prochaska, 1993). Even with very different types of behaviors, the stages of change are similar (Rimer, 1996).

The transtheoretical model:

- Conceptualizes behavior change as an ongoing process, not a single event.
- Suggests that, in changing an existing behavior or adopting a new one, people move through a series of stages.
- People who are not thinking about changing are in the precontemplation stage (❶), those thinking about changing are in the contemplation stage (❷), those planning to change in the near future are in the preparation stage (❸), those in the process of changing are in the action stage (❹), and those trying to maintain a change are in the maintenance stage (❺).
- Suggests that not all people are equally ready to change, and that different intervention strategies will be required depending on the stage of change a person is in.

Most health behaviors must be maintained to have effective preventive health benefits. Although relapse is classically conceived as abstinence violation in a cessation attempt, relapse can occur with a variety of behaviors. For instance, a woman who has passed the date on which she should have had a repeat pap smear can be considered a relapser. A former

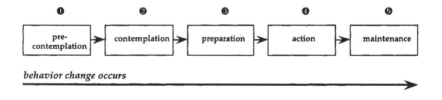

FIG. 3.5.   Transtheoretical model ("stages of change").

bike-helmet wearer who has begun to ride without a helmet is a relapser. Rates of alcohol relapse over time suggest that there may even be different stages of change even within the relapse process. For example, people who are trying to survive the first 2 weeks without alcohol may be in a different relapse prevention stage than those who have been abstinent for a year (Marlatt, 1985).

According to stage theory, individuals at different stages of a behavior process differ from each other in the decisional balance of perceived positive and negative aspects of the new behavior. Perceived benefits refer to the positive components of taking a health action, whereas barriers refer to the negative aspects of a health behavior. Persons who have performed a behavior should have a decisional balance favoring the behavior, people not yet considering the behavior should have a decisional balance reflecting reasons not to perform the behavior, and those contemplating the behavior should have a decisional balance in between. This reflection of decisional balance by stage has been born out in a variety of health behaviors, including mammography screening (Rakowski, Fulton, & Feldman, 1993; Rakowski et al., 1992), smoking cessation (Velicer, DiClemente, Prochaska, & Brandenburg, 1985), and weight loss (O'Connell & Velicer, 1988).

As its title indicates, the model is transtheoretical. Variables and constructs delineated in other theoretical models can be considered across the stages of change. For instance, in a study exploring the interface between beliefs specified by the HBM and the transtheoretical model's stages of change, benefit perception scale scores were higher as subjects approached the Action/Maintenance stage for screening mammography, whereas barrier scores were higher for women who were not considering having a mammogram than among those who were considering or who had undergone the procedure within the recommended time frame (Champion, 1994).

Interventions may target different variables at different stages. Individuals in the precontemplation stage might need information about susceptibility and benefits, contemplators may need messages addressing barriers or enhancing self-efficacy regarding the behavior change, relapsers may need redirected attributions that focus on changeable, controllable causes.

Applications for Behavior Change.   Stage conceptualizations are intervention-oriented. Their premise is that people at different points in the health behavior decision process may behave in different ways, with different factors influencing their behavior, and need different kinds of interventions and information to move them closer to taking preventive health actions (Weinstein, 1988).

For addictive behaviors, the transtheoretical model identifies relapse prevention, or maintenance, as distinctly different from other stages in the behavior change process and suggests that the strategies one uses to, for instance, prevent relapse to drinking should be different from those used to help an alcoholic stop drinking in the first place. Specific recommendations might include building relapse prevention skills such as identifying risk situations and developing coping strategies, reminding of positive outcomes already achieved, and changing expectations about the effects of alcohol use (Marlatt, 1985).

Use of a stage model does not suggest that the process is as important as the behavior itself (e.g., using a seatbelt). However, interventions can be more effective if they consider the process through which individuals decide to change a behavior and what kind of help or information they need at different points in the change process. The person who has worn a bike helmet in the past may need a very different kind of intervention from someone who has never worn a helmet.

Thus, stage models have important implications for tailoring interventions to the stage, or readiness to act, of their intended audience (Rakowski et al., 1992; Weinstein & Sandman, 1992). Prochaska and DiClemente's categorization of smokers by readiness to quit (Prochaska & DiClemente, 1983, 1988) has been found so useful for smoking cessation interventions, the National Cancer Institute has listed a specific target to stage as an essential element of such interventions (Glynn, Boyd, & Gruman, 1990).

AN INTEGRATIVE FRAMEWORK FOR HEALTH PROMOTION

A planning model can facilitate the behavioral diagnosis process and the development of a theoretical framework for intervention. The PRE-CEDE/PROCEED model of health promotion program planning has been applied to a variety of health-related problems in many settings, and provides a useful framework from which to analyze determinants of health-related behaviors (Green & Kreuter, 1991). This model helps identify factors influencing people's decisions and actions and uses that information to develop appropriate programs and behavior-change interventions. The model suggests health behaviors are influenced by three specific types of factors—predisposing, enabling, and reinforcing factors. *Predisposing* factors are those that can facilitate or hinder a person's motivation to change, and include knowledge, attitudes, values, and beliefs. *Enabling* factors are those that can support or hinder a person's efforts to make the desired behavior change, and include skills, resources, and barriers. *Reinforcing* factors are those that can encourage or discourage continuation of the desired behavior, and include any positive or negative feedback received from others following adoption of the behavior.

As an example, consider the problem of parental compliance with immunization recommendation for preschool children. Lack of knowledge about the childhood immunization schedule and the perception that immunization is not important are both factors that would predispose parents not to have children immunized. Lack of time and transportation, as well as health care system factors such as inaccessibility of services, limited clinic hours, and restrictive policies (e.g., immunization by appointment only) are nonenabling factors that function as barriers to immunization compliance. Finally, dissatisfaction with clinic care and prolonged crying by the baby after receiving shots would be a disincentive to immunizing a child, thus reinforcing noncompliance. As illustrated in Fig. 3.6, factors such as these influence whether a child is fully immunized, which, in turn, influences the child's likelihood of contracting preventable childhood illnesses, which, in turn, affects the quality of life for the parents and child. To help interrupt this spiral, the messages used in an immunization promotion campaign would need to address some of these factors that influence immunization compliance (see Fig. 3.6).

Although oversimplified for the purpose of illustration, even this cursory behavioral analysis of childhood immunization reveals some of the factors that might influence parents' compliance with immunization recommendations. If a health promotion campaign sought to promote childhood immunization among parents without assessing or addressing such factors, it would likely fail—whether its approach was theory-based or not. Therefore, the more detailed your understanding of the behavior you are seeking to change and the determinants of that behavior, the less likely it is that your program will miss the mark.

PRECEDE/PROCEED's systematic approach to diagnosing health-related problems facilitates critical thinking about intervention points and strategies. The framework allows for broad analyses on a variety of levels (i.e., individual cognitive, situational, social–environmental). Relevant constructs from most health behavior theories can be conceptualized in the PRECEDE/PROCEED planning model. Thus, use of PRECEDE/PROCEED can expand health promotion planning beyond a single theory or

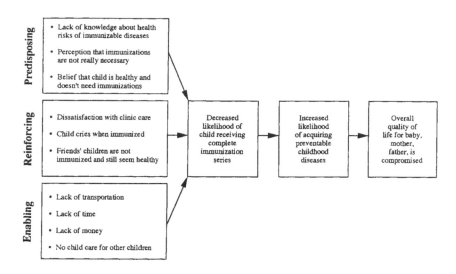

FIG. 3.6.   PRECEDE/PROCEED model and nonadherence with immunization recommendations.

model while still allowing for the rigorous application of theory to practice (Gielen & McDonald, 1996).

Relationship of Health Belief Model to the Planning Model.
Completing the PRECEDE/PROCEED behavioral diagnosis process should generate a list of predisposing, reinforcing, and enabling factors related to the behavior of interest. If HBM variables appear on this list, the intervention program should address them. Consider again the problem of low childhood immunization rates. If a mother believes her child is healthy and therefore does not need any shots (a predisposing factor against immunization), the HBM would likely classify this as low perceived susceptibility. Likewise, the belief that getting an immunizable disease is just a part of growing up (a predisposing factor against immunization) would indicate low perceived severity in HBM terms. Lack of transportation or time (enabling factors against immunization) would be considered an HBM barrier. Figure 3.7 illustrates the relationship between HBM and the PRECEDE/PROCEED planning model.

If factors like perceived susceptibility, severity, barriers, and benefits showed up in a behavioral diagnosis of the target population, the HBM could be useful in guiding an intervention to address the problem of interest. According to HBM, if a person does not perceive some threat from the health problem, taking preventive action is unlikely. Therefore, any predisposing factors relating to perceived susceptibility or perceived severity should be the first to be addressed.

Relationship of Efficacy Theory to the Planning Model.   If the PRE-CEDE/PROCEED behavioral diagnosis process reveals that people do not think they are capable of executing the desired behavioral change or do not believe that making the change will lead to a positive consequence, efficacy theory may be an appropriate model to guide intervention planning. Consider the challenge of promoting safe sex practices among sexually active college students. If students felt they could not easily discuss condom use with a potential partner before having intercourse (a predisposing factor and an example of low self-efficacy for negotiating condom use) or if they believed that talking to partners about condom use would not increase its perceived acceptability (a predisposing factor and an example of negative outcome expectations), it is unlikely such communication would take place.

Relationship of Attribution Theory to the Planning Model.   If the behavioral assessment using the PRECEDE/PROCEED model reveals that internal, uncontrollable, stable, or global attributions about the behavior of interest are common among members of the target population, it may be useful to consider ways that attribution theory can inform your intervention. Consider the problem of relapse among recovering alcoholics. Studies show that as many as 60% of those who try to quit drinking relapse within the first 2 months after quitting, and most people try several times before finally quitting for good (Hunt, Barnett, & Branch, 1971). Marlatt (1985) proposed that one of the factors that influences relapse is the kind of attributions a person makes when first violating his or her newfound abstinence. If the person believes the reason for the drink was because, "I'm weak and have no willpower" (an internal, stable attribution and a reinforcing factor against continuing trying to maintain abstinence), the chances are greater that he or she will continue to drink and revert back to previous drinking behavior.

Relationship of Theory of Reasoned Action to the Planning Model.
If the PRECEDE/PROCEED behavioral diagnosis reveals that the members of the target population have negative attitudes toward the behavior of interest, or believe the behavior to be anormative for members of their social group, the (TRA) could help inform an intervention to address some of these factors. For example, assume a behavioral assessment of physical activity among older adults revealed that most thought there was little benefit from exercise for older people (a negative behavioral belief in TRA, and a predisposing factor against exercise). Furthermore, most of their friends did

not exercise regularly (normative beliefs in TRA, and an enabling or reinforcing factor against exercise), and they did not want to be the only ones doing it (high motivation to comply with inactivity in TRA, and an enabling factor against exercise). According to TRA, this kind of behavioral profile is unlikely to result in intention to do regular physical activity; hence, the likelihood of behavior change occurring would be small. Given such a profile, an intervention should address these factors as recommended by Ajzen and Fishbern (1980). The relationship of TRA to PRECEDE/PROCEED is shown in Fig. 3.7.

Relation of the Transtheoretical Model to the Planning Model.   The PRECEDE/PROCEED educational assessment process is closely related to the Transtheoretical Model's explanation of the behavior-change process. The Transtheoretical Model suggests that people who are not seriously thinking about changing (precontemplators) are unlikely to do so. In the PRECEDE/PROCEED model, these people would not, by definition of stage, have predisposing factors for making the desired behavioral change. Contemplators and those in the preparation stage, on the other hand, are more favorably predisposed to change. Their likelihood of changing is largely determined by the factors that enable or impede their efforts to change. These are enabling factors in the PRECEDE/PRECEDE model. Finally, making and maintaining behavioral changes are the focus of, respectively, the action and maintenance stages of the transtheoretical model. Success in these stages is often determined by the extent to which the behavioral changes already made are reinforced over time. Such factors are considered reinforcing factors in the PRECEDE/PROCEED model. The relationship of the Transtheoretical Model to PRECEDE/PROCEED is shown in Fig 3.7.

Having completed the educational diagnosis in PRECEDE/PROCEED, planners will find that tenets of the transtheoretical model are useful in

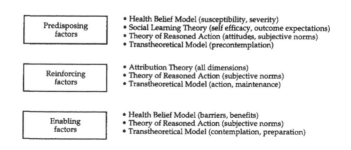

FIG. 3.7.   PRECEDE/PROCEED model's relation to the 5 specified models.

prioritizing behavioral change strategies, regardless of what theoretical models might be applied. For example, for behaviors found to have important predisposing, enabling, and reinforcing factors, it may be important to address the predisposing factors before other factors. Consider the case of smoking cessation. A behavioral cessation plan that focused on overcoming barriers (i.e., enabling factors) without identifying or addressing readiness to change (i.e., a predisposing factor) would be misguided. Overcoming barriers for precontemplators is irrelevant because they are not considering changing. A more appropriate strategy in this case would be to address the predisposing factors first. This will frequently be the case when predisposing, enabling, and reinforcing factors are present for the behavior of interest.

Summary

The PRECEDE/PROCEED planning model can aid health promotion planners in identifying factors influencing health-related behaviors in target populations. Which factors are identified in the behavioral diagnosis will indicate which theory (or combination of theories) should form the framework for a health promotion program. Community or group programs usually target factors influencing the health behaviors of most program recipients. However, interactive computer programs can target health promotion messages to individual recipients.

APPLICATIONS OF THE THEORETICAL FRAMEWORKS FOR INTERACTIVE COMPUTER PROGRAMS

Interactive computer programs provide a unique opportunity to conduct individualized behavioral diagnoses and deliver messages based on specific theoretical constructs that are most important for a particular program user (i.e., his or her predisposing, enabling, and reinforcing factors). Group interventions based on any of the models surveyed in this chapter have not been capable of maximum effectiveness because they cannot specifically target the most important variables for each *individual* recipient. Interactive computer programs are much more able to assess which variables are most important to users and then provide tailored messages addressing those specific variables. The following is a discussion of interactive applications for the models presented in this chapter.

Interactive Applications for the Health Belief Model. According to the HBM, weighing the relative benefits and barriers to taking preventive action is irrelevant (because people will not do it) if there exists no perception of threat. When a threat is perceived, the HBM suggests that the balance of barriers and benefits will determine whether preventive action is taken. Under these circumstances, an interactive computer program might provide ways to overcome nonenabling factors that were barriers to change while reinforcing perceived benefits to change or introducing new benefits the person may not have considered.

A particular HBM variable may be more important for Individual A than for Individual B. One person may have an entire set of misperceptions that work together to inhibit health behavior action, whereas another person may have only one major barrier inhibiting action. Programs can determine a person's strongest, or most important, barrier to the behavior change in question and then provide messages designed to reduce that particular barrier. However, some barriers are more easily reduced via computer program than are others. For instance, barriers resulting from misconceptions may be reduced by persuasive, understandable presentations of accurate information. Other barriers, such as costs of medical screening procedures, may not be so easily dissipated via an interactive program. In such a case, information about insurance coverage or sources of reduced-fee screening could be delineated or the relative weight of potential benefits could be compared with anticipated costs. Benefits already perceived can be reinforced and those not perceived can be communicated.

Interactive Applications for Efficacy Theory. Of the four sources of self-efficacy expectations, interactive computer programs may be especially effective in three. By using skill-building exercises and clearly elucidating the user's success and skill acquisition, self-efficacy can be enhanced through performance accomplishment. By showing a person similar to the user successfully accomplishing a task, computer programs can also boost self-efficacy through vicarious experience. Demographic and psychological information collected from the user can determine that the characteristics of the model, who is portrayed as achieving success in the program, is similar to the user. Verbal persuasion can obviously come through information presented in a program. What kind of information presented, and the way in which it is presented, can be tailored according to user characteristics. Although computer programs may be more limited in using physiological cues, the program can be designed to keep users feeling calm and it can use models to demonstrate how to cope or get back on track when one experiences negative physiological cues.

Interactive computer programs might enhance low perceived self-efficacy by teaching specific behavioral skills and modeling them, and by providing opportunities for the user to rehearse these behaviors in simulated situations. Making the initial behavioral objectives relatively simple and easy to achieve can enhance self-efficacy by giving people early successes to build on, and focusing on the positive aspects of an incomplete performance. Providing reinforcement for users' efforts may also enhance perceived self-efficacy. If outcome expectations are negative, interactive computer programs should clearly demonstrate the relationship between the behavior and outcome and provide opportunities for the user to experience specific outcomes as a result of the decisions he or she has made. It may also be important to conceptualize the outcomes of interest as more immediate and tangible, and not as a long-term health benefit to attain in the distant future.

Interactive Applications for Attribution Theory.   Interactive computer programs can make strong use of attributions (i.e., attributions for causes of slips in abstinence, of emotional responses to the lapse, and of efficacy expectations for future success in similar situations). In short, when people believe the cause for a lapse was a personal weakness that is unchangeable and not controllable, they are more likely to feel guilty and bad about themselves, thereby increasing the negative effect of the abstinence violation, decreasing self-efficacy, and increasing the likelihood of the lapse becoming a full-blown relapse. Reattribution interventions can direct focus toward external causes. For instance, if a recovering alcoholic violates abstinence at a bar, he or she may attribute this lapse to lack of willpower. If this person sees this internal cause as stable and uncontrollable, his or her self-efficacy will be lowered. Lowered self-efficacy will, in turn, decrease this individual's chance of success in future attempts. An interactive reattribution exercise could attempt to focus the attribution on an external cause, such as the situation of being in the bar where other people were drinking and the person himself had been accustomed to drinking in the past (Marlatt & Gorden, 1980). Although the person perceives inability to control his or her drinking in the bar, the intervention could focus on his or her ability to avoid the bar altogether, thus building perceived efficacy for abstinence maintenance.

Interactive Applications for the Theory of Reasoned Action.   An interactive computer program could begin by determining whether the user intends to make a change in one or several specific behaviors. In cases where behavioral changes would reduce the user's risk but the user did not intend to make a change, the program could branch into a more detailed assessment of behavioral beliefs, outcome evaluations, normative beliefs, and motivation to comply. By measuring these variables, the program could determine which components of intention should be targeted to increase the likelihood of changing intention. The computer program could deliver messages designed to enhance expectations for a certain behavioral outcome or the desirability of that outcome. Although a computer program may not change the opinions of salient others, it can help the user to identify and seek out those who endorse the behavior (e.g., health professionals) and attempt to enhance their motivation to comply.

Interactive Applications for the Transtheoretical Model.   Stage models of health behavior change consider that people at different points in the health behavior decision process behave in different ways and that the kinds of interventions and information needed to move people closer to action will vary from stage to stage (Weinstein, 1988). Although mass-produced intervention messages cannot specifically address an individual's stage of behavior change, interactive computer programs are well equipped to do so. Assessing readiness to act, on a continuum of stages, then targeting messages toward an individual's stage of readiness, is a more effective health intervention paradigm (Prochaska, 1994) than plans trying to push everyone—including those who are not close to being ready—toward immediate action. Interactive programs can provide messages and exercises designed to move precontemplators toward consideration of the target behavior change. Further, because the goal of interventions is to move people into the Action/Maintenance stage and keep them there, it is important to identify relapsers and those at risk for relapse.

In this Transtheoretical model, constructs from other models can be differentially applied for persons at different initial stages of consideration or action. For instance, relevant others' opinions (a TRA construct) about adopting a behavior could be emphasized for a contemplator, whereas others' negative opinions about discontinuing the health-related behavior could be emphasized for a person at risk of entering the relapse stage.

SUMMARY

In 1973, Leventhal reported that the ability to tailor messages was one of the most important components of face-to-face communications. More than 20 years later, the programming and production capabilities of microcomputers allow for message tailoring that approximates the advantages of face-to-face tailored communications (Skinner, Siegfried, Kegler, & Strecher, 1993). The enhanced message relevance can result in greater likelihood for behavior change. However, tailored messages can only be effective for encouraging behavior change if they are tailored to specific factors influencing the message recipient's behavior.

Interactive computer programs should consider characteristics of the user, the context in which he or she will need to make the behavior change, the type of behavior change required, and the context in which the program will be used. The PRECEDE/PROCEED planning model and theories of behavior change outlined in this chapter can aid program planners in considering what the program should do. Then, the development team should work together to produce a well-designed and engaging program to meet those objectives.

This chapter has addressed the nature of target health behaviors or health problem, the determinants of the behavior among members of a specified population, and the processes of behavior change, including knowledge and application of a range of theories that address the determinants of the target behavior or health problem. Other issues for consideration in developing programs—interactive computer technologies and communication strategies for creating an effective health promotion program for the population of interest—are addressed in following chapters.

REFERENCES

Ajzen, I., & Fishbein, M. (1980). *Understanding attitudes and predicting social behavior.* Englewood Cliffs, NJ: Prentice-Hall.

Annis, H. M., & David, C. S. (1988). Assessment of expectancies. In D. M. Donovan, & B. A. Marlatt (Eds.), *Assessment of addictive behaviors.* New York: Guilford.

Bandura, A. (1977a). *Social learning theory.* New Jersey: Prentice-Hall.

Bandura, A. (1977b). Toward a unifying theory of behavior change. *Psychological Review, 84*(2), 91–215.

Bandura, A. (1982a). The self and mechanisms of agency. In J. Suls (Ed.), *Psychological perspectives on the self (Vol. 1).* Hillsdale, NJ: Lawrence Erlbaum Associates.

Bandura, A. (1982b). Self-efficacy mechanism in human agency. *American Psychologist, 37*(2), 122–147.

Baranowski, T., Perry, C., & Parcel G. (1996). How individuals, environments, and health behavior interact: Social cognitive theory. In K. Glanz, F. Lewis, & B. K. Rimer (Eds.), *Health behavior and health education* (3rd ed, pp. 153–178). San Francisco, CA: Jossey-Bass.

Champion, V. L. (1994). Belief about breast cancer and mammography by behavioral stage. *Oncology Nursing Forum, 21*(6), 1009–1014.

Chaney, E. J., O'Leary, M. R., & Marlatt, G. A. (1978). Skill training with alcoholics. *Journal of Consulting and Clinical Psychology, 46,* 1092–1104

Curry, S. J., & Emmonds, K. M. (1994). Theoretical models for predicting and improving compliance with breast cancer screening. *Annals of Behavioral Medicine, 16*(4), 302–316

DiClemente, C. C., & Hughes, S. (1990). Stages of change profiles in outpatient alcoholism treatment. *Journal of Substance Abuse, 2,* 217–235.

Gielen, A. C., & McDonald, E. (1996). Applying theories of health behavior to program planning in health promotion. In K. Glanz, F. M. Lewis, & B. K. Rimer (Eds.), *Health behavior and health education: Theory, research, and practice* (2nd ed., pp. 359–383) San Francisco, CA: Jossey Bass.

Glanz, K., Patterson, R. E., Kristal, A. R., DiClemente, C. C., Heimendinger, J., Linnan, L., & McLerran, D. F. (1994). Stage of change in adopting healthy diets: Fat, fiber, and correlates of nutrient intake. *Health Education Quarterly, 21*(4), 499–519.

Green, L. W., & Kreuter, M. W. (1991). *Health promotion planning: An educational and environmental approach* (2nd ed.). Mountain View, CA: Mayfield Publishing.

Grimley, D. M., Riley, G. E., Bellis, J. M., & Prochaska, J. O. (1993). Assessing the stages of change and decision-making for contraceptive use for the prevention of pregnancy, sexually transmitted diseases, and acquired immunodeficiency syndrome. *Health Education Quarterly, 20*(4), 455–70.

Glynn, T. J., Boyd, G. M., & Gruman, J. C. (1990). Essential elements of self-help/minimal intervention strategies for smoking cessation. *Health Education Quarterly, 17*(3), 329–345.

Janz, N. K., & Becker, M. H. (1984). The health belief model: A decade later. *Health Education Quarterly, 11*(1), 1–47.

Hochbaum, G. M. (1958). *Public participation in medical screening programs: A sociopsychological study.* Public Health Service, PHS Publication no. 572. Washington, DC: U.S. Government Printing Office.

Hospers, H. J., Kok, G., & Strecher, V. J. (1990). Attributions for previous failures and subsequent outcomes in a weight reduction program. *Health Education Quarterly, 17*(4),409–415

Hunt, W. A., Barnett, L. W., & Branch L. G. (1971). Relapse rates in addiction programs. *Journal of Clinical Psychology, 27,* 455–456.

Kirscht, J. P., & Rosenstock, I. M. (1979). Patients' problems in following recommendations of health experts. In G. Stone (Ed.), *Health Psychology.* San Francisco: Jossey-Bass.

Leventhal, H. (1973). Changing attitudes and habits to reduce risk factors in chronic disease. *American Journal of Cardiology, 31,* 571–580.

Lierman, L. M., Young, H. M., Kasprzyk, D., & Benoliel, J. Q. (1990). Predicting breast self-examination using the theory of reasoned action. *Nursing Research, 9*(2), 97–101.

Marlatt, G. A. (1980). Relapse prevention: Theoretical rationale and overview of the model. In G. A. Marlatt & J. R. Gordon (Eds.), *Relapse prevention: maintenance strategies in the treatment of addictive behaviors* (pp. 3–70) New York: Guilford.

Marlatt, G. A., & Gordon, J. R. (1980). Determinants of relapse: Implications for the maintenance of behavior change. In P. O. Davidson & S. M. Davidson (Eds.), *Behavioral medicine: Changing health lifestyles.* New York: Brunner Mazel.

Montano, D. E., & Taplin, S. H. (1991). A test of an expanded theory of reasoned action to predict mammography participation. *Social Science and Medicine, 32*(6), 733–741.

O'Connell, D., & Velicer, W. F. (1988). A decisional balance measure for weight loss. *International Journal of Addictions, 23,* 729–750.

Prochaska, J. O., & DiClemente, C. C. (1983). Stages and processes of self-change of smoking: Toward an integrative model of change. *Journal of Consulting and Clinical Psychology, 51,* 390–395.

Prochaska, J. O., & DiClemente, C. C. (1988). Measuring process of change: Applications to the cessation of smoking. *Journal of Consulting and Clinical Psychology, 56,* 520–528.

Prochaska, J. O. (1994, April). *Staging a revolution.* Keynote address: Fourth anniversary Meeting of the Society of Behavioral Medicine, Boston, MA.

Rakowski, W., Fulton, J. P., & Feldman, J. P. (1993). Women's decision making about mammography: A replication of the relationship between stages of adoption and decisional balance. *Health Psychology, 12*(3), 209–214.

Rakowski, W., Jube, C. E., Marcus, B. H., Prochaska, J. O., Velicer, W. F., & Abrams, D. B. (1992). Assessing elements of women's decisions about mammography. *Health Psychology, 11*(2), 111–118.

Rimer, B. K. (1996). Perspectives on intrapersonal theories in health education and health behavior. In K. Glanz, F. M. Lewis, & B. K. Rimer. (Eds.), *Health behavior and health education* (pp. 139–148). San Francisco: Jossey-Bass.

Rogers, E. M. (1962). *Diffusion of innovations.* New York: The Free Press.

Rosenstock, I. M. (1991). The health belief model: Explaining health behavior through expectancies. In K. Glanz, F. M. Lewis, & B. K. Rimer (Eds.), *Health behavior and health education* (pp. 39–58). San Francisco: Jossey-Bass.

Rosenstock, I. L., Strecher, V. J., & Becker, M. H. (1988). Social learning theory and the health belief model. *Health Education Quarterly, 15*(2), 175–83.

Rossi, J. S. (1989). Exploring behavioral approaches to UV risk reduction. In A. Moshell & L. U. Blankenbaker (Eds.), *Sunlight, ultraviolet radiation and the skin* (pp. 91–93). Bethesda, MD: National Institutes of Health.

Skinner, C. S., Siegfried, J. C., Kegler, M. C., & Strecher, V. J. (1993). The potential for computers in patient education. *Patient Education and Counseling, 22*(1), 27–34.

Strecher, V. J., DeVellis, B. M., Becker, M. H., & Rosenstock I. M. (1986). The role of self-efficacy in achieving health behavior change. *Health Education Quarterly, 13*(1), 73–79.

Timko, D. (1987). Seeking medical care for a breast cancer symptom: Determinants of intentions to engage in prompt or delay behavior. *Health Psychology, 6*(4), 305–328.

Strecher, V. J., & Rosenstock, I. M. (1996). The health belief model. In K. Glanz, F. M. Lewis, & B. K. Rimer (Eds.), *Health behavior and health education: Theory, research, and practice* (2nd ed., pp. 41–59). San Francisco, CA: Jossey-Bass.

Velicer, W. F., DiClemente, C. C., Prochaska, J. O., & Brandenburg, N. (1985). A decisional balance measure for assessing and predicting smoking status. *Journal of Personality and Social Psychology, 48,* 1279–1289.

Wallston, B. S., & Wallston, K. A. (1984). Social psychological models of health behavior: An examination and integration. In A. Baum & J. Singer (Eds.), *Handbook of psychology and health* (Vol. 4, pp. 23–53). Mahwah, NJ: Lawrence Erlbaum Associates.

Weiner, B. (1986). *An attributional theory of motivation and emotion.* New York: Springer-Verlag.

Weinstein, N. D. (1988). The precaution adoption process. *Health Psychology, 7*(4), 355–386.

Weinstein, N. D., & Sandman, P. M. (1992). A model of the precaution adoption process: Evidence from home radon testing. *Health Psychology, 11*(3), 170–180.

4

Interactive Environments for Promoting Health

Timothy Manning
Texas A&M University Health Science Center

The previous chapters have provided theoretical frameworks for looking at a variety of processes that can influence the effectiveness of a health-promotion enterprise (e.g., educational outcomes of the user–media–message interaction, factors affecting the adoption of health behaviors). This chapter examines some of these issues from a different perspective. It examines different approaches to health promotion by considering the roles played by health promoters in relation to their audience, and the ways computing does and can facilitate these roles. It is argued that conceptualizing computing as a medium (instead of a tool) lends itself particularly well to approaching health promotion as the empowerment of individuals in support of their own initiatives to integrate health behaviors into their lifestyles. Such a perspective can best support those health promotion relationships in which the individual is more independent in making health decisions and choosing health behaviors. An approach referred to here as *health integration* creates a central role for a highly differentiated, self-responsible individual and a facilitative role for the health promoter.

HEALTH PROMOTION

There has been considerable progress in the field of health promotion over the last two decades. A more comprehensive definition of health has been developed, a greater diversity of health enhancement objectives has been identified, and a more complex view of the individual has emerged along with a better understanding of differences in how individuals value health. With each step in this evolution we have created new types of relationships between the role of the health promoter and the role of the individual.

At one time, health was a relatively monolithic concept. Health meant the absence of medical pathology, a definition that could be applied to the health of anyone. Although we continue to understand health to mean medical wellness, social, psychological, and economic aspects of health also have become important (Draper, Griffiths, Dennis, & Popay, 1980, Smith, 1979). Health encompasses well-being, functional capability, quality of life, and fitness, in addition to the health of the body (Downie, Fyfe, & Tannahill, 1990). This broader meaning of health has created an overlap of the goals of prevention, treatment, health education, and public health and brought this overlap into the legitimate jurisdiction of health promotion. Whereas the dimensions of health to which we attend have increased (e.g., medical, social, familial), so too have the health factors we consider relevant. Occupation, education, lifestyle, and social and physical environment all have been productively scrutinized for their effect on health (National Center for Health Statistics, 1995). In addition, the recent emphasis on medical outcomes and the reorientation of the health care system toward managed care has further elevated the importance of health promotion, so that the field now stands as a relevant and dynamic part of the mission of health care organizations (see, for example, chapters 11 and 12, this volume).

Two key changes can be perceived in these developments. One is that health must be considered to be a complex mixture of many factors that varies from individual to individual. The second is that health must arise from an individual's personal choices in seeking to establish a balance between those factors that affect health and one's own values, needs, wants, and goals. Thus, the responsibility of individuals for their own health has become more relevant as health promotion finds its role in empowering individuals to make appropriate personal health choices (e.g., diet, exercise) in addition to promoting behaviors believed to enhance health (smoking cessation, cancer screening).

ROLES OF THE HEALTH PROMOTER

The strategies used in enhancing individual health have progressed and multiplied to reflect multiple roles for the health promoter. Health promotion initiatives cast the health promoter in four roles: messenger, educator, marketer, and facilitator. Each has its appropriate place in the overall health promotion arsenal, but they do vary in sophistication in how they conceive the individual. Although these roles often occur in combination, by examining them individually we can identify how interactive technologies may be used to support each role.

The health promoter as messenger is one who delivers information and, for the most part, conceives the individual in simple terms. Frequently, a single message is delivered to a largely undifferentiated audience. A good example is the surgeon general's warning on cigarette and alcohol products. There are two assumptions in this role. First, health behavior choices are assumed to be entirely intellectual. Noncognitive understandings such as social, emotional, and behavioral are largely ignored. There is also an assumption that the individual is already motivated to strive toward good health. That is, one who does not engage in the targeted health behavior simply lacks the information. Once that information is provided, the behavior will be changed. This role is quite appropriate for certain objectives, for instance in explaining the use of medication (e.g., dosage and frequency). The simple exposure to information can be effective (Zajonc, 1968) particularly if delivered by a prestigious communicator (Aronson, Turner, & Carlsmith, 1963). However, patients do not always comply even if they have the information. Is this because the patient is not motivated, has difficulty understanding or accepting the information, or is averse to using the medication?

The health promoter as educator takes information delivery to a somewhat higher level of sophistication. Health information is still provided, but understanding is not assumed to occur simply on presentation of the information. Instead, the information is provided in such a way as to employ the cognitive learning processes of the receiver of the information. New concepts are broken down into manageable chunks and connections are made between new concepts and existing knowledge. Techniques are used to organize and elaborate the information so it is easier to comprehend and remember (Anderson, 1990; Miller, 1956). Opportunities may be provided to rehearse learned knowledge (Resnick, 1985) and to embed new skills within familiar contexts (Rumelhart & Ortony, 1977). By making use of learning theories, the education approach can be much more effective than simple information delivery, at least with respect to recall and retention of information. As in health promoter as messenger, this approach primarily relies on cognitive processes involved in learning and skill acquisition to accomplish the health promotion objective. However, the motivation to learn and produce the target behavior continues to be assumed, or at least is not considered. For instance, an industrial safety training program may be effective in increasing knowledge about safety but may not adequately respond to the variation in beliefs of individual workers about the likelihood of injury.

The health promoter as marketer recognizes that the individuals may not be motivated to make lifestyle changes or to expend energy in the acquisition of new health knowledge, at least in the way the health promoter deems desirable. Hence, in addition to the provision of information, a primary objective of the marketing approach is to persuade and motivate the individual to adopt or maintain the target health behavior. This approach may make use of learning theory, but does not simply assume that the user is a willing partner. The benefits of adopting the target behavior are emphasized. Fear and other emotional appeals may be employed to motivate the audience. Methods of more easily changing the health behavior may be suggested. However, the goal is still to obtain targeted health behaviors. Examples of health promotion as marketing would include public service announcements for the "Say No to Drugs" campaign or for promoting child immunization programs.

The diversity of health concepts, health beliefs, values, and the greater focus on the individual as sharing responsibility for health suggest a fourth role, health promoter as facilitator. When one attempts to facilitate individual initiatives in adjustment to medical exigencies and the adoption of new health behaviors, the objective is health integration. The objective of the health promoter in this approach is to support the individual's integration of new knowledge about health and healthy behaviors into his or her current knowledge systems and lifestyle (Riley, 1986).

Instead of force feeding a particular idea or behavior, this approach relies on the individual's interest and desire to incorporate the information into new understandings that are particularly useful and appropriate for his or her personal circumstances. Thus, this approach respects the differences in individuals and supports their ability to make their own choices concerning their health. It equips them with information resources and tools needed to make such choices. Its goal is the empowerment of the individual rather than the elicitation of a specific health behavior. For example, a patient with diabetes will integrate new behaviors and skills and form new health beliefs in a personal way that corresponds to the person's characteristics. What is important in this process is not the adoption of a specific behavior, but an overall adaptation to the disease that may occur in a variety of ways across different individuals.

COMPUTING AS A MEDIUM

Before examining the role of computing in each of these roles, it may be useful to examine two perspectives (metaphors) on computing and show how one in particular is well suited for achieving the goals of health promoter as facilitator in health integration.

The computer is often referred to as a tool. This has been a very useful metaphor, but it does not fully capture the breadth, depth, and possibility of computing applications. A more powerful metaphor conceives computing as a new kind of medium. To understand computing as a medium is to redirect our attention from the physical objects involved (tools) to the broader ground on which we operate (medium). Certainly the computer is a salient and ever-present part of projects involving computing, but one cannot understand computing by understanding the computer, just as it would be impossible to understand television by studying a television receiver. Television would not be possible without the technology, but an understanding of the technology is neither sufficient nor necessary for producing an understanding of television as a medium. The salience of the hardware associated with a new technology is initially so strong that we tend to focus our attention on the devices rather than the new functionality the devices make possible.

Our first attempts to use a new medium are based on what we do understand; that is, previous media and their associated technologies. The motion-picture camera was first understood as a kind of displaced eye and used in order to see things that had occurred at another time and place. Thus, many early movies were patterned on the stage play. The drama unfolded in front of a fixed camera that merely recorded the action from the best seat in the house. The content of the motion picture became the medium of theater as this new medium seemed to encompass and supersede the play presented on the stage. McLuhan wrote that "the 'content' of any medium is always another medium." (1964, p. 23). We do what we know how to do. We put plays on movies and books on computers, then, as more experience is gained, distinct qualities of the new media emerge.

Our first understanding of computers was in terms of mechanization and automation. In the paradigm of the industrial age, we thought of the computer as machine. We used it as a sort of mill for churning numbers, for processing data. Later images of the computer characterized it in human terms. We anthropomorphized the computer, programming it to be a psychotherapist, educator, or manager (Weizenbaum, 1976). A research program was developed to create artificial intelligence. In each of these conceptualizations, there is the device and there is the way the world works in light of its existence. With respect to computing, what is most important is not so much the device as it is the way things work in this new media space.

When we began to use traditional tools, we set up a media space in which our ability to manipulate the physical environment was extended. Instead of adapting to the environment, we began to change it. This is the watershed development that led to industrialization. It is the idea that we can construct the physical world to suit our needs and desires. This was well depicted in a scene from the motion picture *2001: A Space Odyssey*. In this scene, a Neanderthal gains the insight that a bone can be used as a club. In an outburst of delighted power, the Neanderthal tosses the bone into the air. The camera follows it up as the sky darkens into a black, star-studded sky and the bone dissolves into a space station. It is not that we discovered the club. It is that we came to understand a powerful way to control our physical surroundings. The existence of tools, or rather the idea of tools, creates a new medium in which we shape the environment to our purposes.

Information technologies, including computing, point to another watershed insight. The idea of the tool is that physical objects can be formed and arranged to manipulate the physical environment, but the insight of information technologies is that objects need not be physical in order to achieve the desired result. Symbolic abstractions are free of many of the restrictions of physical objects. If information is the object of interest, our manipulative power is unaffected by most physical constraints. The computer, or the computing idea, creates a new medium in which we have far greater power than in the physical environment. The relevant object in this new environment is information.

INFORMATION AS AN ENVIRONMENT

Equipped with the tool metaphor, one would describe using the telephone as sending messages back and forth. Using the medium concept, one would describe the shared experience of the audio space set up by the telephone. There is a sense of presence in using the telephone that cannot be explained by the tool metaphor. If you are speaking to someone on the telephone, in what sense are you not with the person?

The concept of information as an environment broadens the tool metaphor to include the participant, a diversity of tools and media (within the medium of computing), and a system of rules. The computer becomes a window into this environment in which tools and other objects operate within their respective media (e.g., narration, text, music, motion picture), interacting, interpenetrating, evoking each other, subordinating each other, creating the potential for experience.

An environment is alive, dynamic, responsive. The users' actions on it are consequential. Responses to user actions may be either consistent and understandable or surprising and evocative. An environment has an ecology. It is not necessarily static in the absence of user action, but may evolve on its own. It may change with the actions of others. It is complex and rich with possibility. Most important, it is experiential (Laurel, 1986). It feels like something. The user feels a sense of presence or perhaps feels immersed in the space (Steuer, 1992).

A tool is used to perform a specific task with a specific outcome. An environment creates the potential for experience where the specific outcome is not precisely predictable. It depends on the interests, intentions, and other aspects of the user and how those are supported by the environment. Of course, an environment may facilitate certain kinds of action over others but, importantly, it also provides context and situatedness.

The greatest potential for computing is not in the delivery of information. It is in the creation of environments that are rich with the potential for independent action and discovery. The construction of such an environment focuses on the user (Nilan, 1992). The environment must be different enough from the real environment to provide and expand on the possibilities for experience and similar enough to the real environment so the user understands what is possible in it, what actions are appropriate.

If we take the computer to be a tool, we restrict ourselves too much from what computing has to offer. Using computing as a medium, we can construct rich information environments in which we engage the individual in the fullest sense, as someone with cognitive abilities, beliefs, attitudes, values, and as one who expresses considerable responsibility for health choices. This is a fitting approach to helping individuals integrate health concepts and behaviors into their unique worlds of experience.

INTERACTIVE MEDIA IN HEALTH INTEGRATION

In Fig. 4.1 we see how computing orientations may be applied to the health promotion roles discussed previously. The roles of messenger, educator, and marketer all typically have in common the goals of a specific targeted health behavior. The health promoter as a facilitator is different in that it seeks to help the individual make personal choices about his or her health, or to follow a loosely defined learning or recovery path. For the first three roles, the computer is most often viewed as a tool in that it provides a means of providing information or service to achieve the targeted outcome. Computing viewed as a medium is most compatible with the role of health promoter as facilitator.

	Information Delivery to Undifferentiated Individuals	Education of Individuals with Cognitive Processes	Persuasion of Individuals with Beliefs, Attitudes, and Values	Empowerment of Individuals in their Health Initiatives
Health Promotion Approaches	Messenger	Educator	Marketer	Facilitator
Computing Orientation	Information Resource	Processing and Practice Opportunities	Structured Presentation	Discovery Learning Decision Support Path Finding

FIG. 4.1   Computing orientations in health promotion approaches.

Computing and the Messenger

Computers, databases, and networks can be powerful ways of getting information to people. The health promoter as a messenger selects the information appropriate to the individual and uses the computer as a delivery device. If the user is provided considerable independence in navigating the information, the computer becomes a tool of the user rather than a tool of the health promoter. Given sophisticated navigational tools, some aspects of an environment become apparent. The conception of information as not only text and numbers but also as images, music, speech, movies, and simulations creates more potential to provide the user with aspects of an environment. In this case the health promoter is no longer acting simply as a messenger.

Computing and the Educator

Learning theories suggest many ways computers can be used in an educational approach to health promotion. Computers can be tools to help the developer and the user organize information. For example, a common strategy is to create programs so that the user controls the pace of the presentation. A self-paced program that also offers options for which topics to view in what order can help the user connect new knowledge to existing knowledge, which in turn can increase comprehension and retention. As the software tools become more general, the role of the user becomes more independent and the medium aspect of computing becomes more apparent. The user will play a larger role in the learning process. To date, however, most computing applications for health education, although they may be described as learning environments, deal only with cognitively relevant information and processes like comprehension and retention. A good example would be a program on diabetes management that provides information on nutrition and diet, then tests recall.

Computing and the Marketer

In the use of computing to persuade individuals to adopt health behaviors, structured presentations are the most prevalent. Generally, program developers want to make sure all points, arguments, and appeals are presented to the users. Thus, the presentations tend to be linear and the user is offered few options for exploration. Even if options are presented, the program is constructed so that it ensures no information is excluded. For instance, a program advocating the use of birth control pills may force the user to view all possible effects in order to insure informed consent. A persuasive approach does not, however, restrict one from using computing as a medium. By providing structured experiences, more than one view can be emphasized. This would make more use of computing as a medium than a simple presentation. However, as the user's degrees of freedom increase, the ability to present a single point of view probably declines.

Computing and the Facilitator

The role of health advocate as facilitator places the individual at the highest level of initiative potential and offers the most potential for using computing as a medium. In this setting, the health promoter is responsible for creating the possibilities for discovery, for providing resources for health decisions, and possibly for offering opportunities for interaction with others. Facilitation is appropriate to situations in which something is understood about the desired form of the experience, but the decision the individual is to arrive at or the specific behavior to be adopted is largely undetermined, and it will depend on the individual's personal preferences and circumstances.

For instance, consider a person dealing with a serious health threat. It has been said that we live our lives as a drama in which we play the role of the central character (Taylor, 1989). We generally understand what happens in terms of how it affects us, and we tend to interpret events based on the role we see for ourselves. When a patient receives a diagnosis of a life-threatening disease such as cancer, the development of that drama is thrown into uncertainty and the role of the self is likewise thrown into question. Much of the effort of the patient is now directed toward rebuilding their world to account for and accommodate this new situation. This person will attempt to find meaning for the event, establish mastery over the situation, and improve self-esteem (Taylor, 1983; Taylor, Lichtman, & Wood, 1984). What the health facilitator needs to do is provide the opportunity for the patient to follow his or her own path in making an adjustment that is personal.

So what kind of information environment would we create to help these patients along this path to adjustment? The environment need not be representational of the real world as such, although it does have to be meaningful and understandable. A cancer patient will still have a real-world experience with health care professionals, other cancer patients, family, and friends. The computer experience need not duplicate that. However, it may allow them to participate in modeling exercises that help them identify a new role for themselves. For example, it may allow them to hear the experiences of other cancer patients or to contribute their own experiences. It could allow them to participate in simulated radiation sessions or chemotherapy treatments to familiarize them with the processes and to prepare them to make decisions about treatment. It could facilitate the making of social comparisons through interaction with patients, recordings of patients, or simulations of patients.

Consider another example—diabetes management. People with diabetes will have to adopt many new behaviors if life is to be extended and medical complications avoided. They will change how they eat and exercise. They will need and seek support from others. They will learn problem-solving skills to cope with events and complications (e.g., unexpected exertion, hypoglycemic reactions) that before their disease would have been trivial. But the specific behaviors and the balance of those behaviors can vary. For instance, a large variety of reasonable diet choices and exercise plans can achieve the same result. People need emotional support in different degrees and have different ways of finding it. Some may suffer from depression with the disease or endure unusual family pressures. Their adjustment has a knowledge aspect, a skill aspect, an emotional aspect, and a social aspect, all of which are important in addressing these problems. If a program is to allow different individuals to sort out how they want to change their life to accommodate their disease, it must allow for many different solutions.

CONCLUSION

For someone making the journey of adjustment to a health threat or considering a healthier lifestyle, the environmental model promises to place the locus of initiation and benefit with the patient. It promises to support general, rather than specific, objectives and thus allow for different paths to adjustment. And it promises to make affective, as well as cognitive, factors more relevant in the use of computing for health promotion. Conceiving information as an environment provides an alternative view that is more appropriate to those health promotion efforts in which the personal aspects of the individual are most relevant. Where the individual plays such a central role, where there is a reasonable expectation for individual initiative, where the goal is the encouragement of personal health objectives, the concept of health integration may provide an effective approach to health promotion.

REFERENCES

Anderson, J. R. (1990). *Cognitive psychology and its implications* (3rd ed). San Francisco, CA: Freeman.

Aronson, E., Turner, J., & Carlsmith, J. M. (1963). Communicator credibility and communication discrepancy. *Journal of Abnormal and Social Psychology, 67*, 31–36.

Downie, R. S., Fyfe, C., & Tannahill, A. (1990). *Health promotion: Models and values*. New York: Oxford University Press.

Draper, P., Griffiths, J., Dennis, J., & Popay, J. (1980). Three types of health education. *British Medical Journal, 281*, 493–495.

Laurel, B. (1986). Interface as mimesis. In D. Norman & S. Draper (Eds.), *User centered system design: New perspectives on human–computer interaction* (pp. 67–86). Hillsdale, NJ: Lawrence Erlbaum Associates.

McLuhan, M. (1964). *Understanding media: The extensions of man.* New York: McGraw–Hill.

Miller, G. A. (1956). The magical number seven, plus or minus two: Some limits on our capacity for processing information. *Psychological Review, 63*, 81–97.

National Center for Health Statistics. (1995). *Healthy people 2000 review, 1994.* Hyattsville, MD: Public Health Service.

Nilan, M. S. (1992). Cognitive space: Using virtual reality for large information resource management problems. *Journal of Communication, 42*, 115–135.

Resnick, L. B. (1985). Cognition and instruction: Recent theories of human competence. In B. L. Hammonds (Ed.), *Psychology and learning: The master lecture series* (Vol. 4, pp. 127–186). Washington, DC: American Psychological Association.

Riley, M. (1986). User understanding. In D. Norman & S. Draper (Eds.), *User centered system design: New perspectives on human–computer interaction* (pp. 157–190). Hillsdale, NJ: Lawrence Erlbaum Associates.

Rumelhart, D. E., & Ortony, A. (1977). The representation of knowledge in memory. In R. C. Anderson, R. J. Spiro, & W. E. Montague (Eds.), *Schooling and the acquisition of knowledge* (pp. 99–135). Hillsdale, NJ: Lawrence Erlbaum Associates.

Smith, E. A. (1979). Health education and the National Health Service. In I. Sutherland (Ed.), *Health education: Perspectives and choices* (pp. 93–112). London: George Allen and Unwin.

Steuer, J. S. (1992). Defining virtual reality: Dimensions determining telepresence. *Journal of Communication, 42*, 73–93.

Taylor, S. E. (1983). Adjustment to threatening events: A theory of cognitive adaptation, *American Psychologist, 38*, 1161–1173.

Taylor, S. E., Lichtman, R. R., Wood, J. V. (1984). Attributions, beliefs about control, and adjustment to breast cancer. *Journal of Personality and Social Psychology, 46,* 489–502.

Taylor, S. E. (1989). *Positive illusions: Creative self–deception and the healthy mind.* New York: Basic Books.

Weizenbaum, J. (1976). *Computer power and human reason: From judgment to calculation.* San Francisco, CA: Freeman.

Zajonc, R. B. (1968). Attitudinal effects of mere exposure. *Journal of Personality and Social Psychology, Monograph Supplement, 9,* 1–27.

II

USING INTERACTIVE TECHNOLOGY TO IMPROVE HEALTH

5

Aiding Those Facing Health Crises: The Experience of the CHESS Project

Robert P. Hawkins
Suzanne Pingree
David H. Gustafson
Eric W. Boberg
Earl Bricker
Fiona McTavish
Meg Wise
Betta Owens
University of Wisconsin–Madison

We describe an interactive computer-based system that provides information, help, and support in a variety of packages to individuals facing a health-care crisis. But before discussing the Comprehensive Health Enhancement Support System (CHESS) project, it may be useful to raise some issues and caveats that apply to all of us. In the book and movie, *Field of Dreams*, an Iowa farmer plows under a field of corn and builds a complete baseball field because voices whisper to him, "If you build it, they will come." In the story, a dream team of baseball legends does come and play, justifying the farmer's seemingly senseless actions.

The caveat here is that new interactive technologies also can be seductive. The lure of the hardware—the flash of "look what we can do"—can lead us to ill-considered and ultimately useless system development. Sometimes it seems as though the developers have heard voices saying, "If you build it, they will use it," and pursue system construction with little regard for the user or outcomes. Such systems provoke gee-whiz reactions, but may not be used or, if used, may not accomplish anything useful.

In particular, there are two underlying issues, the first (as already implied) having to do with technology and the second with information. The first issue is illustrated by the rapid adoption of the Information Highway metaphor. Here the voices seem to be saying, "If you connect them, they will be healthier (or wiser, more creative, better citizens, etc.)." But the highway metaphor (or its alternative, the information repository) misleads us: We tend to think about building the highway and assume that the traffic and benefits will follow. Instead, we need to understand where people want or need to go, and then build vehicles (or tools) that actually take them there. For example, if we want to have patients who are informed, responsible, and proactive, or to have doctors and medical systems that can integrate information effectively and efficiently, or to connect individuals and groups into communities, these goals require more attention to the vehicles than to the information highway. The applications that make such goals possible are the key bridges between humans and hardware, and it is the analysis of goals (what is needed) and strategies (how to achieve it) that is most crucial. This is exactly the situation that public communication campaigns in general have faced up to in recent years by analyzing the routes to desired behavior changes and constructing a mix of messages, structural changes, and interpersonal supports, each targeted (separately or in interactions) to specific subgoals of the overall goal (Salmon, 1989; Strecher, McEvoy-DeVellis, Becker, & Rosenstock, 1986).

In addition, however, a second problem is common among applications of new technologies, and here the voices seem to be saying, "If you give them information, they will use it." That is, in many cases, the implicit assumption is that access to information is the key requirement for success. But conceiving information as something discrete, a thing to be accessed, is itself part of the problem, as scholars of communication and information have known for years, at least in principle (Dervin, 1980; Rappoport, 1953; Schramm, 1955). Although we as senders or receivers may perceive discrete messages or documents or facts to be information, they are instead merely the stimulus for the construction of information—the reduction of uncertainty about the state of the world. Thus, it is the needs and perceptions of individuals that define whether a fact is informative or not; access to the facts may not even be relevant. What this leads to, however, is a conclusion that meshes very well with that of the previous paragraph in its focus on end goals and users rather than on the facts and messages: How facts are presented (and, in particular, whether people are able to make use of them) will often be more important than which facts are presented.

Given these caveats and the alternatives presented, we propose a short set of principles particularly relevant to developing new technology applications relevant to health.

1. Build tools. Some developers have used the metaphor of developing a communication toolbox, and we think this metaphor is worth taking seriously. Tools are the thing that humans use, and tools are shaped both by the task and to fit the human hand. The central tasks for humans coping with and using complex data to extract useful information are discovering ways to recognize relevance to problems and discovering ways to manipulate and use both the data and the resulting information. So rather than focus on the technology or the bits of information per se, the key to successful applications is conceiving them as things people will use to address their problems.

The task of turning connections into tools requires analysis of human needs and capabilities to start with, and it should take into account current habits and expectations people have of information and of communication media. For example, data about cancer may be much more useful if it is reshaped to respond to specific common questions or if it can be integrated into a decision tool. The electronic patient record will be an overwhelming string of facts unless it can respond to queries and produce only the relevant information (perhaps it should also analyze and summarize trends). And if people have come to expect screens to be simpler and more entertaining than the printed page (an unproven relationship often blamed on television), we will have to either oblige or retrain them.

2. Train and be trained. Both developers and users of new technologies for health applications will probably need to alter their behaviors. Users will have to be motivated and trained to use information systems because these systems will often offer capabilities and flexibilities to which users are unaccustomed. In particular, the systems will often require the user to provide the information on which tailoring is based and to make choices at many points. As a result, they will often require more commitment and activity from users. Similarly, developers—both the technologists and the content experts—will have to be trained to recognize needs of the general public and also in ways to respond to those needs. Quite often, the expert's view of a problem is significantly different than that of the general public. What the expert assumes about system capabilities and the meaning of direction for using the program may be different from those of the general public.

3. Weave into daily life. Although some health applications can probably attract users on their own, the economic and training costs of these technologies mean that substantial user motivation may be needed to overcome barriers of cost and habit. One way to minimize such barriers is to weave health-technology applications into everyday life and other applications as much as possible. The principle is to bring users into routine or even

incidental contact with the application, to build awareness and familiarity, encourage experimentation, and connect the benefits of the application to other needs and goals. Any technology or interface risks getting minimal use if it stands on its own. But if health applications share technologies with other (preferably everyday) tasks, they will be perceived as easier to use and be more likely to come to mind when needed. Better still, where health applications can be integrated with everyday applications (e.g., nutrition analysis of a grocery list), they can become familiar and routine technologies and be integrated with services for the other aspects.

4. Start small. Implicit but obvious in this is that applications development is complex, requiring attention to a variety of psychological, social, and economic issues, among others. Less obvious but equally important is the sense that real-world utilization and dissemination of applications is very much at boot-strap stage. That is, even a psychologically sensitive and wonderful system may be difficult to disseminate because individual and social expectations and the infrastructure are not yet ready for it. Or a superb technology may leap ahead of our ability to create applications that really use its potentials. It is a commonplace in the study of communication technologies that we tend to understand a new technology as just an extension or improvement of the old, and only slowly recognize the ways in which it is qualitatively different (the automobile as a horseless carriage, television as radio with pictures, interactive cable as just more choices).

What this means for the developer, unfortunately, is an environment full of uncertainty and one that often punishes assumptions. The moral may well be: Start simple and move up with perpetual prototypes. Our use of technologies will be better suited to both development and dissemination if we make our technologies scalable. That is, we need to be able to try something out in a small and simple way and then build it on up if it works. In the near term, perpetual prototyping is a development model of constant and ongoing formative evaluation. The alternative—build it big and complete the first time—is just too prone to failure, given our current understanding of information, users, and the potentials of technologies.

5. It's ok to be captivating. Commercial products and communications spend large amounts of money being flashy, easy to use, and visually appealing. We should not feel that these characteristics are beneath us. In part, this is a corollary to weaving into everyday life. Style is a part of everyday life in consumer society, and products that look and feel awkward and unfinished are simply less likely to be used.

DEVELOPING CHESS

These considerations, especially the first, third, and fourth, figured large in the initial thinking that led to the CHESS system. As a first step, we had to decide what sort of health problems should best be targeted, and we were guided first by a judgment that some health applications would initially be more easily woven into everyday life than others. Given current uneven and rapidly changing development of relevant technologies, and a spotty mix of systems using them, applications designed to serve everyday and prevention health needs of general populations seemed problematic for initial development. Such applications cannot assume high levels of motivation and focused interest in their content. Although the lack of such motivation and interest may not be a problem in the future when the general population routinely uses a technology (perhaps Internet- based) for a wide variety of everyday services, and thus can be drawn to the general health applications through casual contact or a variety of links, we judged that even the most useful stand-alone prevention application might be used only rarely or by only a very few individuals.

Therefore, our initial development efforts have all focused on health or life crises of some sort for two reasons. First, we believed that individuals facing a health crisis would be more motivated to seek out and use a system than those without such pressing needs. Second, because we realized that any system would have to start small or risk being too shallow to be credible, the focus of a particular health crisis provides justification for preparing a narrow range of material in considerable depth and flexibility. That is, instead of covering a wide variety of health topics superficially, we believed we could be more effective and achieve both use and usefulness by covering specific topics systematically. Thus, although we recognized that any particular topic would have relevance only to a minority of the population, we sought to prepare a system that would be highly relevant and very useful to that minority, so that they would make heavy use of it and also, we hoped, would gain substantial benefits. From the limited populations of several quite specific health crises, however, we hoped to expand rapidly because we expected that the tools we developed would be easily translatable to a wide variety of other specific crises.

Deciding on a particular health crisis to address involved a cycle of assessing needs, identifying how those needs could be met, and, finally, judging whether computer-based technologies could facilitate addressing the needs. Each of these steps has been addressed elsewhere, but is briefly summarized here.

Needs assessment must be sensitive both to what people want and what they need and must recognize that the two may not be the same (Gustafson et al., 1993). Although it involves surveys, interviews, and focus groups of people who are facing or have faced the health crisis, it also draws on health care professionals who routinely treat the crisis, and other experts, in addition to the scientific literature on the disease or problem. The second step involves an analysis of theory and past empirical research on potential paths to meet that need, whether through learning, behavior change, identification and provision of social support, and so forth. We drew on a wide range of theory and research on aspects of crisis (Aguilera, 1990; Moos & Schaeffer, 1984) and behavior change (Bandura, 1977; Fishbein & Ajzen, 1980; Strecher et al., 1986). Finally, based on our previous experience designing and evaluating health education software (Bosworth, Gustafson, & Hawkins, 1994; Hawkins, Gustafson, Chewning, Bosworth, & Day, 1985) and our analyses of what computer-assisted technologies can and cannot be expected to do (Bosworth & Gustafson, 1991; Gustafson, Bosworth, Chewning, Hawkins, 1987; Hawkins & Pingree, 1989), some tools will be a much better match to a given need than others.

Take breast cancer as an example. Patients often need extensive information to help them understand and evaluate their diagnoses, treatment options, and so forth; guidance and tools to make choices about treatment (perhaps relating treatment options to both medical and nonmedical outcomes) or plan behavior change strategies (e.g., diet changes); training in seeking out and making best use of medical and other services; strategies to deal with family and friends; and substantial ongoing emotional and practical support, which often can best be obtained from fellow breast cancer patients. This analysis of needs and strategies led directly to the development of the main services or tools of the CHESS Breast Cancer module (Gustafson et al., 1993).

It is also worth noting that people facing health or life crises currently face a variety of barriers in accessing the information and support they need, and computer-based systems are well-suited to overcome many of these barriers. For example, the information and support people need is often currently difficult or inconvenient to get, with libraries requiring travel and conforming to specific hours that may not match when questions arise, and those containing specialized information relevant to a health crisis often the most difficult to access. An in-home computer system can be immediately at hand at any time of the day.

Another barrier typically facing people seeking health information or support is that much material may be incomprehensible due to unexplained medical terminology. Medical literature is not reader-friendly, and clinicians often lack the training to communicate in ways patients can understand. A well-designed computer system can both attend to vocabulary issues and break complex information into more manageable units.

Furthermore, many people find hospitals, clinics, and even libraries threatening, for any of multiple reasons. Besides the impersonal, public space and the association with the health crisis itself, clinicians themselves operate under increasing time pressure and may seem rushed, intimidating, and uncaring. Using a computer at home places the material and support in a safe and familiar setting. In addition, the pace and direction of the interaction in those formal settings (with the exception of libraries) is often not in the user's control, whereas computer-based systems allow users to proceed at their own pace, dealing with issues how, when, and in what order they wish.

Because of underlying similarities in the needs assessments for several initial CHESS topics, CHESS was designed as a shell of about a dozen services or tools in three general categories of *information, decision and planning support*, and *social support*. Thus, the same basic computer programming of each service incorporated different specific content (or could be omitted altogether), depending on the needs associated with a particular crisis. The services are each designed to meet a particular need in a given way, so that in some cases, two different services may contain overlapping information presented in a different context or in greater or lesser depth. In addition, the classification of services as *information, decision support*, or *social support* is somewhat arbitrary because several clearly fall in at least two categories.

At present, CHESS has six completed systems: breast cancer, HIV infection, sexual assault, adult children of alcoholics, academic crisis, and stress management. Modules on substance abuse, Alzheimers, and heart disease are underway, and others are being planned. Although the system uses color and graphics to provide an attractive and very easy-to-use interface (computer novices get a half-hour of training to become comfortable using the system), the system will run on any 286-or-higher machine equipped with a modem and at least 20 MB of free hard-disk space. At present, the entire program resides on the user's hard disk, but the modem provides electronic mail connections to experts or to other people facing the same crisis.

Information Components

Questions and Answers is a compilation of brief answers to many (typically several hundred) common questions about the topic. *Instant Library* is a collection of complete articles drawn from scientific journals, newsletters, brochures, pamphlets, and the popular press. These provide greater depth and integration, but less specificity than Questions and Answers.

Getting Help/Support is a tutorial on how to find and evaluate needed services (such as health care providers and social service agencies), and how to get the most benefit from interactions with service providers. For example, the tutorial in some CHESS modules suggests agendas and even specific wording for discussions with providers. *Ask an Expert* allows users to ask specific questions of content experts, who provide specific answers within 48 hours. Experts have access to the individual user's *Profile*, so that their answers can take individual circumstances into account. (These exchanges are often intended or perceived as social support as well.) *Dictionary* provides easy-to-understand definitions of technical or medical terms.

Social Support

Personal Stories are personal accounts of individuals' experiences with the health crisis. Written in the first person by professional journalists after in-depth interviews, these present a wide range of background and experiences. Users can read a basic story in each case and can also select any of a number of expansions to pursue a particular interest (e.g., how the person dealt with a brusque doctor). *Discussion Groups* provide a nonthreatening place to communicate anonymously, at any time of day or night, with others facing the same health or life crisis. Users share information, experiences, hopes and fears, give and receive support, and offer different perspectives on common issues. Because a given discussion group typically contains 5 to 25 members, participants log in to read much more often than they write messages. Because of this, they not only find responses to their own messages, but also encounter discussions of topics they have not yet considered, leading them to a broader picture of their situation and their options. (Both Personal Stories and Discussion Groups also convey considerable information).

Analysis and Planning Services

Health Charts, in several modules, periodically assess the severity of symptoms and offer feedback on courses of action (e.g., call a physician today vs. monitor for several days). In other modules, Health Charts allow an individual to track various aspects of quality of life over time. *Profile* is a list of questions designed to describe the background and crisis characteristics of the individual. Completed during the first CHESS use session (and updatable later), the Profile is then available in *Ask an Expert* to allow considerable personalization of answers. *Assessment* uses subjective Bayesian models and interactive interviewing to assess a user's situation and behaviors, and provides feedback that identifies problematic behaviors (e.g., the risks of transmitting the HIV virus, or identifying life domains that are contributing to academic difficulties; Gustafson, 1987). *Decisions and Conflicts* uses utility theory to help people think through difficult decisions such as what surgery to have, whether to discuss their HIV status with coworkers, or what sort of professional help to seek. For common decisions in each module, users can elect to read about treatment options (including relevant excerpts from *Personal Stories*) and about criteria that can be used in making the decision. They can also generate their own options and criteria for any decision. They then select options they will consider, and weight and rate criteria for each option. CHESS does not tell users what to do, but users can choose to see how the computer used their input to predict which alternative they would choose (Sainfort, Gustafson, Bosworth, & Hawkins, 1990). *Action Plan* uses decision and change theories to help users think through how they will implement a decision (e.g., change to a low-fat diet despite habits and family pressures) by analyzing their strengths and weaknesses, supports and barriers, predicting their likelihood of success, and suggesting ways they can improve their prospects.

USE OF THE CHESS SYSTEM

Most of the CHESS modules have been pilot-tested with a variety of specially recruited samples, often contacted through cooperating physicians, social service organizations, university living units or counseling services, corporate work places, and so on. In addition, the HIV, Adult Children of Alcoholics, and Breast Cancer modules have undergone more formal testing, including random assignment to CHESS or non-CHESS conditions. A number of these tests have already been reported elsewhere, so this section will briefly summarize salient findings from various studies.

CHESS Is Used Heavily.　All field tests have shown repeated and heavy use, but the most complete, impressive results come from the longer-term tests in which computers have been left in homes 2, 3, or 6 months. For example, HIV-infected users in a field experiment used CHESS an average of more than once per day over a 3-month period (see Table 5.1). Although use was, of course, heaviest during the first few weeks of access, one cohort allowed longer access was still using CHESS an average of three times per week after 6 months of continuous access. (We should note that use was not perfectly normally distributed: Medians were about two thirds of the means, due to a tail of extremely heavy users, and about 10% of those with CHESS computers used them less than once a week.) Clearly, the system was not a resource that could be easily used up; some needs are simply ongoing and, in a long-term disease such as HIV infection, others continually appear (Boberg et al., 1995).

Table 5.1 also breaks CHESS use into the nine main services. Here, as in most CHESS trials, usage of *Discussion Group* is dominant both in number of uses and in total time of use, typically accounting for over 70% of all uses and an even higher percentage of total computer time. However, because total use of CHESS is so high overall, the information and analysis tools still see substantial use, generally more than 2 times per week. It is also worth noting that all services are used at least once by the majority of participants, with most sampled by almost all participants. The analysis services are clearly used less often than other portions of CHESS, although one should also note that such tools are more likely to have their benefit used up in one or two uses than the diverse information services or the constantly changing Discussion Groups.

TABLE 5.1
Use of CHESS Servises by HIV-Infected Individuals
During a 3-Month Experimental Trial

CHESS Service	% of Subjects Using Service	Average Number of Uses per Subject	Average Time of Each Use (min)
Total CHESS Use	—	138.0	17.4
Social Support Services			
Discussion Group	100	104.0	20.6
Personal Stories	96	5.2	7.7
Information Services			
Instant Library	96	7.4	10.1
Q & A	98	5.6	4.8
Ask an Expert	87	8.2	6.8
Getting Help/Support	80	2.1	5.0
Analysis Services			
Decision Analysis	78	2.2	8.1
Action Plan	67	1.5	8.3
Risk Assessment	53	1.4	5.0

Interesting differences are evident in the pilot tests of CHESS with breast cancer patients, who use the computer slightly less often per week, but still are heavy users. However, some women get done with the system and stop using it after 3 to 6 months, whereas others strongly resist returning the computer at the end of the 6-month experimental period. Exit interviews suggest that breast cancer was a limited, episodic illness for some women who wanted to put the experience behind them after completing treatment, whereas others experience the disease as having considerable ongoing consequences and readjustments, for which CHESS continued to be useful.

CHESS use occurs at all hours of the day and night, with heaviest usage typically between 5 and 11 p.m. In the HIV trial, more than one third of all uses occurred between 9 p.m. and 7 a.m., a time when few, if any, other sources of information and support are available. Among women with breast cancer, such late-night use accounts for well over 40% of all CHESS use (Gustafson et al., 1993).

The Disadvantaged Use CHESS as Much or More.

An important concern with new technologies (and historically with almost every print-based or information-rich technology) is that groups typically disadvantaged in society will have less access to the information. Even if access is equalized, however, there is still reason to worry that those with less education, the poor, minorities, or others less experienced with a technology (and for computer-based systems, this may include women) will be less likely to use a new technology or will get less benefit from it. We were worried that this might be so with CHESS, despite our attempts to make it easy to use and reduce language demands. However, a detailed analysis in the HIV study showed that differences between demographic subgroups were small, and in most cases, the system was used more by groups ordinarily expected to use these technologies less. In particular, women and minorities made more use of several information tools, and minorities and those with less education used the decision and analysis tools more (see Table 5.2), even though those tools are the most complex and demand the most from the user of any tool in the system (Pingree et al., 1996).

Similarly, a pilot study gave CHESS to eight African-American women with breast cancer receiving their medical care through Cook County Hospital. Only half of the women had completed high school and all were on public assistance, living in impoverished inner-city neighborhoods of Chicago. These women used CHESS slightly more than the predominantly

TABLE 5.2
Education and Ethnic Group Differences in the Use of CHESS Services

	Race		Education	
	Minority	Majority	Low	High
Total Number Uses	99.45	117.00	121.36	102.32
Total Minutes	1800.00	2065.00	2054.00	2067.00
Q & A	6.82*	4.87	5.61	4.62
Instant Library	8.95*	6.41	7.41	6.06
Getting Help/Support	2.55*	1.79	2.22***	1.15
Personal Stories	5.64	4.49	5.20*	3.53
Risk Assessment	1.36	1.14	.96	1.65
Decision Aid	3.18**	1.76	2.13	1.85
Action Plan	2.05**	1.20	1.52	1.03
Discussion Group	61.73	85.26	88.58	69.50
Ask An Expert	6.55	6.74	5.25**	9.74
N	22	85	69	34

Note. Minority refers to African-American and Hispanic.
*$p < .10$. **$p < .05$. ***$p < .01$.

White, well-educated, and middle-class women in a prior pilot study in Madison. And again, they made relatively more use of information and analysis tools than the middle-class women.

Two things are probably at work in these results. First, the motivation provided by a life-threatening illness may overcome habits or inclinations not to use information media, given CHESS' simplicity. Second, traditional sources of information and support may be more difficult to access or understand (and because most U.S. HIV-infected individuals are male, HIV-infected women may have some difficulties obtaining information and support), thus making the need for something like CHESS most pressing among the disadvantaged (Pingree et al., 1996). At any rate, it is highly encouraging that CHESS was heavily used by diverse types of people and even used slightly more by those who usually have the least access to information.

EFFECTS OF CHESS

Although users in the pilot tests and field experiments have given CHESS uniformly high ratings on usefulness and ease of use, and many have spontaneously posted testimonials in *Discussion Group* (such as those printed at the end of this chapter), the most convincing evidence of effect comes from experiments involving random assignment to CHESS or non-CHESS conditions.

Although two studies are underway (one with breast cancer and one providing long-term CHESS access for patients with advanced HIV disease), the only such experiment yet completed involved 204 HIV-infected people from Madison or Milwaukee (21 of whom were women, 33 minorities), about half of whom got CHESS computers while the other half merely responded to the same survey instruments. All respondents were paid for completion of each survey and only 10% failed to complete the study (more than half of those due to death). Subjects were recruited in three cohorts, the first having CHESS for 6 months whereas the other two cohorts used a 3-month experimental period to increase the number of participants. Surveys were administered at pretest, 2 months and 5 months, with the first cohort also completing a 9-month survey.

Improved Quality of Life. The basic hypothesis that people who were given CHESS would improve more (or worsen less) than the control group on eight quality-of-life indicators (see appendix) could be tested uniformly for the entire sample by comparing pretest levels with the 2-month posttest, when all members of the treatment group still had CHESS. The 5- and 9-month posttests examine effects of longer implementation and whether effects persisted after CHESS was withdrawn. For the first cohort, the 5-month posttest was still during implementation. Because the second and third cohorts had CHESS for only 3 months instead of 6, their 5-month posttest examines the persistence of effects after a 3-month implementation. And the 9-month posttest examines persistence of effects after a 6-month implementation.

After 2 months, CHESS users reported improved cognitive functioning, more social support, and a more active life whereas controls stayed steady or got worse on each of these variables. CHESS users also reported greater improvement than controls did in actively participating in their health care, and decreased levels of negative emotions, whereas controls stayed the same. There were no significant differences between the groups for depression, physical functioning, or reported level of energy.

At 5 months, when the experimental group of the first cohort still had CHESS in their homes, three of these five effects (social support, active life, and participation in health care) were still significant, despite the smaller sample size, and the other two effects were still of the same direction and magnitude (power is quite low with only one third of the total sample).

For people who had CHESS for 3 months (the second and third cohorts), none of the benefits noted at 2 months remained at 5 months, and this must be taken at face value, because the two cohorts together provide adequate power to reject the null hypothesis.

However, for the cohort that kept CHESS for 6 months, two effects remained significant (social support, participation in health care) 3 months after CHESS was removed. For the other quality of life dimensions that had been significant while CHESS was in the home, differences were in the same direction and of equal or greater magnitude than while CHESS was still in the home. Figure 5.1. illustrates the overall reported quality of life for the first-cohort CHESS and control groups across the 9 months of the study.

Following Up on Treatment.   In a small-scale field experiment pilot testing the module for adult children of alcoholics, subjects were assigned to receive either, neither, or both CHESS and group therapy. Given the small sample, quality of life and most other effects were merely suggestive, as was expected from a pilot. But whereas many group-only subjects dropped out of their therapy group (and the study), therapy group members who also had a CHESS computer almost never dropped out of group therapy during the 10 weeks of the study.

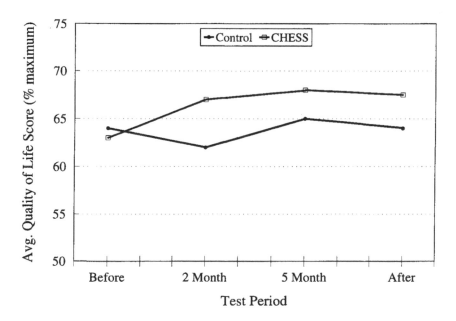

FIG. 5.1.   Average quality of life score before, during, and after CHESS implementation.

More Efficient use of Health Services. In the HIV field experiment, CHESS did not significantly affect number of outpatient visits to providers, although the number of phone contacts increased for the CHESS group compared to controls. However, the CHESS group reported spending significantly less time with providers during visits. One of the doctors whose clinic treated study participants (both CHESS users and controls) told us he could tell which patients had CHESS. The CHESS users came in with prepared lists of questions, had often done background readings, had clear and reasonable expectations of clinic staff, and in general, were ready to be efficient with their own and clinic time. Patients from the control group were more likely to panic, needed information repeated, and in general, took more time.

The one exception to the reduced-time finding is that average time in ER visits was slightly, though not significantly, higher for the experimental group. Looking only at those who reported ER visits, the time spent in the emergency room when a visit occurred stayed relatively constant throughout the study for the control group whereas for the experimental group it increased. But the average number of visits to emergency rooms declined in the experimental group, whereas it stayed relatively constant in the control group. This suggests that the experimental group may have become less likely to use the ER for minor illnesses, and instead was using it only to deal with significant problems.

Respondents also reported whether they had been hospitalized, and if so, for how long. Even in an HIV-infected population, hospitalizations are relatively rare (only about 40 among these 204 individuals over 6 to 9 months), meaning that analyses are very sensitive to outliers. At pretest, those assigned to begin using CHESS reported slightly longer ($p < 10$) average length of recent hospitalizations than did controls among those who had been admitted to hospitals. Therefore, subsequent evaluations took the pretest differences into account through covariance.

The probability of admission increased for both groups (as one might expect for patients living with a progressively debilitating disease), much more for the control group than the CHESS group (71% vs. 29%). Although average length of stay per visit was initially lower for the control group (as noted earlier), it increased sharply (64%), whereas the experimental group declined (23%) so that it was significantly less than that of the control group.

Cost Savings. Aside from statistically significant differences in service utilization, which are admittedly difficult to document given the enormous variability in these measures, there is also the very concrete question of impact on health care costs. Recognizing the potential instability of some of these results, we can still take them as best estimates and monetize them using local figures for various types of charges (i.e., $5 per phone call, visits ranging from $30 for alternate care to $70 for specialists and $90 for ER, $1485 per day HIV hospitalization). For outpatient services based on the number of visits, the before–after reductions were $33 per month for controls and $46 per month for the experimental group. This slight difference is probably an underestimate of savings given the significantly shorter office visits by the CHESS users, which are not accounted for in the monetization.

The average monthly hospital cost per research subject (all subjects, not just the few with hospitalizations) was calculated for each condition by multiplying total days of hospitalization by the cost per day of stay for AIDS care in the Madison area ($1485), and then dividing by the number of subjects. As shown in Fig. 5.2, the experimental group costs stayed fairly constant whereas the control group costs increased (284% during and 263% after, compared to pretest).

CHESS thus appears to have achieved substantial savings, but estimating the amount saved requires one to deal with the large pretest differences. If the changes from pretest are taken literally, there was an enormous benefit

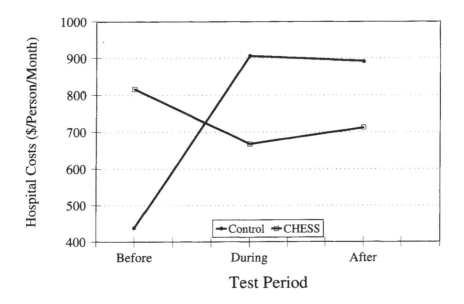

FIG. 5.2.   Estimated hospital costs per research participants per month.

of CHESS, with costs increasing for controls while they held constant for experimentals; CHESS produced a relative savings of $568 per subject per month during implementation and $497 afterward. However, if, despite random assignment, the pretest differences are taken to be unrepresentative and likely to produce subsequent regression to the mean, one could instead use the total-sample pretest average of $467 per month as the baseline for comparing later values for both experimentals and controls. Given this assumption, the differences are instead a savings of $258 per person per month during implementation and $187 per person per month afterward. These smaller amounts still represent 55% and 40% savings from baseline hospitalization costs.

These savings probably come from timely interventions in opportunistic infections. CHESS for HIV/AIDS contains considerable information about symptoms, progression, and treatment of opportunistic infections. In addition, users fill out a *Health Chart* once a week that alerts them to symptoms by asking about them and also allows the users to track their own progress week to week. We believe that the cost savings reported here are largely due to CHESS users seeking treatment at the appropriate time.

For example, pneumocystis carinii pneumonia (PCP) can be misattributed as a lingering cold until symptoms worsen dramatically. At that point, treatment is likely to entail a lengthy, intensive, and expensive hospital stay. In contrast, an individual alert to the symptoms and timing can seek treatment while the infection is much less serious; hospitalization may still be required, but it is likely to be much shorter and less expensive, and the whole episode will be less debilitating. A current field experiment is pursuing this explanation by following sicker CHESS users and controls for a year and monitoring their clinic records as well as their self-reports.

SCENARIOS

These effects are not surprising for people who have used CHESS, but may not be obvious for those who read about it for the first time. The following scenarios are hypothetical, but are built around anecdotal comments we have received from users and physicians.

David and Andrew are both HIV-positive men in their 30's. Both have CD4 counts below 200 (they have AIDS) but neither has yet been hospitalized for an HIV-related illness. David has a CHESS computer in his home; Andrew does not.

Less Time at Clinic Visits

During a regular visit to his HIV clinic, Andrew's doctor notices a small purple lesion on Andrew's back. He tells Andrew that it may be Kaposi's

Sarcoma (KS) and that he will need to have a biopsy. Andrew panics. All he knows about KS is that it means AIDS and can be very disfiguring. He's seen pictures of AIDS patients with advanced KS on TV. He has lots of questions for the doctor: "Does this mean I'm going to die soon? How fast will it spread? Will I look like a freak? How can it be treated?" The doctor not only has to give medical advice but has to do some psychological intervention as well, calming Andrew's fears and reassuring him. This takes a lot of extra time, and the entire clinic schedule is delayed. Furthermore, Andrew's memory for some of the answers he heard while panicked is spotty; he has to call the clinic and ask a nurse some of the same questions the next day.

David is also given a suspected diagnosis of KS during a regular clinic visit. He has read a little about KS in the *Questions and Answers* and *Instant Library*. He knows that it is not immediately life-threatening, that it often spreads slowly, and that treatments are available. Because he knows he has a ready source of information and support in CHESS, he asks his doctor just a few questions, and gets an appointment for a biopsy. David goes home and logs on to CHESS. He reads all of the *Instant Library* material on KS. He reads *Personal Stories* of people who have lived with KS. He leaves a message in *Discussion Group* asking others about their experiences and coping strategies. While in *Discussion Group*, he notices a message in *News Items* on a new experimental treatment for KS. He calls his doctor, who doesn't know much about the treatment, but agrees to find out more about it. David leaves a message in *Ask an Expert* requesting more information on the new treatment. He gets a response back within a day. David also finds the number for Project Inform (a community-based treatment information agency) in the *Referral Directory*. He calls them and finds out more about the new treatment. David also gets feedback on the new treatment from someone in *Discussion Group* who has used it. When he has his biopsy, David and his doctor discuss the pros and cons of the new treatment, and come to a mutual agreement to begin with the standard treatments and see how the KS responds before trying something new.

Reduced Length of Hospitalizations

Andrew develops a dry cough, which persists for a couple of weeks. He assumes it is just a cold. He also notices he is getting fatigued and short of breath more easily, but figures that goes with the territory of having AIDS. Suddenly the symptoms get worse, and he has trouble breathing. A friend drives him to the emergency room. He is admitted to the hospital with a diagnosis of PCP and intravenous Bactrim treatment is begun. However, the infection has taken too strong a hold and it keeps getting worse. A couple of days later, he has to be moved into intensive care and placed on a respirator. After 10 days in ICU, Andrew is taken off the respirator and moved back into a regular ward. He is discharged a week or so after that, much improved, but with a hospital stay of almost 3 weeks, costing in excess of $150,000.

David also develops a dry cough. Because he has read articles in the *Instant Library* and *Questions and Answers* on specific illnesses and has read messages in the *Health Issues Discussion Group,* he knows this could be a warning sign of PCP. He checks with the Symptom Assessment program, which suggests, because he has no other symptoms, that he monitor the cough for a couple of days. At the end of that time, the cough has not gone away, and he is starting to feel a little short of breath. He calls his doctor, who suggests he go to urgent care immediately. The attending physician in urgent care makes a presumptive diagnosis of PCP and admits David to the hospital. Intravenous Bactrim is administered, and because the infection has been caught early, David responds well to treatment. After 3 days, IV treatment is discontinued, and David is discharged to recover at home, where he continues on oral Bactrim. David's body has been stressed much less than Andrew's, and his hospital stay cost less than $5,000.

DISCUSSION

The CHESS experience suggests that computer-based systems directed at health care consumers have the potential to combine the economies of scale of mass media dissemination of rare professional expertise with the depth and specificity of individual information seeking, and also add social support and decision tools. Programs can be written to allow the user to control the pace of the interaction and the choice of subject while providing context and the experts' point of view on issues and decisions. By using the computer as a communication device to connect the individual to experts and to others facing similar situations, human contact can be enhanced even while providing anonymity in dealing with sensitive issues. By making decision tools accessible and usable, computer programs can improve the quality of decision making and planning, and provide many other potential benefits discussed elsewhere in more detail (McDonald & Blum, 1992).

The success of CHESS in research settings now opens questions about dissemination. CHESS has shown itself to be useful and, better still, the various studies have amply demonstrated that people will use it. But as with all better mousetraps, finding a way to deliver the system to all who may need it (for any of the various problems it now does or soon will cover) is by no means trivial.

One issue that may be less serious than it appears is cost. A system that relies on individual computers in homes may appear limited to a tiny, affluent fraction of the population. However, at least for AIDS, the apparent monthly cost savings on hospitalizations are easily large enough to fund distribution by an agency (e.g., an HMO or clinic) that purchases computers (amortizing initial equipment costs over 5 years), operates the system, and provides patients access for several months before moving the computer to another patient. Such organizations are also ideally situated to identify people facing health crises and quickly provide assistance. If equipment can be donated or older computers retrieved from closets, or if some staff time is donated (such as that to support *Ask an Expert* or *Discussion Group*), then the net savings can be even more dramatic.

The more serious difficulties are organizational and psychological. Put simply, CHESS (and other CHESS-like systems) is not yet the way things are done in the health care system. In part, this is a matter of specific attitudes and expectations about health. Disseminating something like CHESS through an HMO or clinic requires active support from physicians and associated clinical staff at least as much as from financial analysts. Although many doctors are eager to have better-informed, responsible patients who take up less of their time, so much so that they are willing to recommend CHESS to their patients (94% of breast surgeons and oncologists participated in a recent 5-county trial), others are doubtful of patients' capacities to make medical decisions or may even feel threatened by activist patients. Some physicians have been concerned, despite the research evidence to the contrary, that a computer-based system could be beyond the capacities of some patients; others have felt that some patients are already overwhelmed by their disease and its associated decisions and could not handle an additional source of information. Either of these concerns can mean that some patients are never offered CHESS. More generally, even if physicians do not oppose making such a system available to their patients, they may see it as simply patient education, and thus less their responsibility than other pressing, purely medical, matters.

Similar but less entrenched biases may operate from the perspective of the potential user as well. Some may not want to take a more active role in their own health. Others, although not opposed to doing so, may still not have imagined the possibility and be slow to seize an opportunity. Some may find the notion of a computer providing them assistance hard to believe.

Others probably are uncomfortable with computers or believe they will be unable to operate one. Reasons like these are probably behind the 20% to 30% who typically decline CHESS computers when offered, a surprisingly low figure.

But beyond this, CHESS and similar systems exist in a partial vacuum because they are still few and far between and not yet woven into everyday life. As more ways of getting health information and support become available and cover more topics, both doctors and patients will become familiar not just with individual systems, but with the general concept of computer-assisted support for individuals. When patients expect to seek information, make decisions, and get social support (and doctors expect that they will do so), then dissemination will accelerate rapidly.

The model of HMO or clinic dissemination is in action now, with a research consortium of a half-dozen major HMOs providing one or more CHESS modules to their patients. In addition, the University of Wisconsin–Madison has licensed a private company, Health Companion Systems, to sell CHESS modules to institutions or individuals. A current CHESS project, sponsored by the Health Care Financing Administration (HCFA) tests the feasibility of making CHESS available to all Medicare-eligible breast cancer patients within a 5-county area to see what level of physician and patient acceptance can be achieved with free access to such a system.

A future system will probably translate such programs onto whatever information highway reaches into all households through interactive television, thus sharing the hardware cost with many other kinds of software and providing near-universal access. At that point, having such social and health services available along with entertainment and commercial services would also qualify them as woven into everyday life and would enable widespread use. At that point, nations may well consider whether public support should be employed to allow universal access to all the benefits of the information highway along with its services.

Finally, if CHESS is indeed a tool for individuals users rather than we health campaigners, it seems fitting to let several CHESS users have the last word with these unsolicited comments placed in the HIV module *Discussion Group*.

Thank you for giving me the computer. I would be lost without it. I hate to think of the day when it gets removed. I haven't been out since July, and it has kept me sane with everything going on with my partner. J. and I isolated ourselves for nine years because of our fear of our HIV status. I now don't have many friends, but I am now gaining more, thanks to CHESS.

D.K.

I ask more and more intelligent questions when I see my doctor. I feel more in charge of my health care. I have met people who are in the same situation I am in. This has helped. It feels good to share things with others, things that will help them. [CHESS] has allowed me to give of myself in a way that otherwise would not have been possible.

M.D.

There is a genuine sense of caring amongst us that I knew we all had, but just were never given this kind of access to. Selfless acts in a time when we are all fighting for our very existence seems like a fairly significant amount of caring to me. I am proud to have been associated with all of this and all of you.

Brooks

APPENDIX

Quality of Life Measures

Four subscales were taken from the Medical Outcomes Survey, short form (MOS; Stewart, Hays, & Ware, 1989). The first of these asked whether and for how long the respondent had experienced limitations on six increasingly demanding forms of physical activity, ranging from eating or dressing to lifting heavy objects or running (Chronbach alpha .87).

The second MOS subscale, cognitive functioning, assessed how often the respondent experienced four cognitive difficulties: with reasoning and solving problems, forgetting things, keeping attention on an activity, and doing activities involving concentration (.91).

The third MOS subscale, energy, assessed how often the respondent felt full of pep, had enough energy to do what they wanted to do, felt worn out, and felt tired (the latter two reversed; alpha .85).

The fourth MOS subscale, depression, assessed how often the respondent felt weighed down, discouraged, despair, or fear because of their health problems (.90).

Although we asked the MOS emotion subscale, our own 13-item scale of negative emotions had better reliability (.90) and was more independent of depression and other indices. Using a 5-point scale, respondents reported how often during the past month their HIV/AIDS had made them feel tense, angry, worried, frustrated, insecure, stressed, bored, fearful, sad, indifferent, helplessness, and pity.

Activity was created for the HIV study. Questions asked for agreement or disagreement that, "there's something I have to do every minute of the day," and, "I have lots of free time" (.88).

Social Support was developed by the authors for previous research (Bosworth, Gustafson, & Hawkins, 1993). Respondents agreed or disagreed on a 5-point scale with three positively and three negatively worded items about their perceptions of having friends for emotional support, help, and caring (.83).

Participation in Health Care was developed for this study. Respondents agreed or disagreed with four statements dealing with ability to discuss treatments with provider, feel in control of health care, have increased my knowledge and preparations before visits, and have been assertive with health care providers. (.80)

REFERENCES

Aguilera, D. (1990). *Crisis intervention: Theory and methodology.* St. Louis, MO: Mosby.

Bandura, A. (1977). Self-efficacy: Toward a unifying theory of behavioral change. *Psychological Review, 84,* 191–215.

Boberg, E. W., Gustafson, D. H., Hawkins, R. P., Chan, C-L., Bricker, E., Pingree, S., & Behre, H. (1995). Development, acceptance, and use patterns of a computer-based education and social support system for people living with AIDS/HIV infection. *Computers in Human Behavior, 11*(2), 289–311.

Bosworth, K., Gustafson, D., & Hawkins, R. (1994). The BARN system: Use and impact of adolescent health education via computer. *Computers in Human Behavior, 10*(4), 467–482.

Bosworth, K., & Gustafson, D. H. (1991). CHESS: Providing decision support for reducing health risk behavior and improving access to health services. *Interfaces, 21,* 93–104.

Dervin, B. (1980). Communication gaps and inequities: Moving toward a reconceptualization. In B. Dervin & M. Voigt (Eds.), *Progress in communication sciences,* (Vol. 2, pp. 73–112). Norwood, NJ: Ablex.

Fishbein M., & Ajzen I. (1980). *Understanding attitudes and predicting social behavior.* Englewood Cliffs, NJ: Prentice-Hall.

Gustafson, D. H. (1987). Health risk appraisal, its roles in health services research. *Health Service Research, 22*(4), 453–465.

Gustafson, D. H., Bosworth, K., Chewning, B., & Hawkins, R. P. (1987). Computer-based health promotion: Combining technological advances with problem-solving techniques to effect successful health behavior changes. *Annual Review of Public Health, 8,* 387–415.

Gustafson, D. H., Wise, M., McTavish, F., Taylor, J., Wolberg, W., Stewart, J., Smalley, R., & Bosworth, K. (1993). Development and pilot evaluation of a computer-based support system for women with breast cancer. *Journal of Psychosocial Oncology, 11*(4), 69–93.

Hawkins, R. P., Gustafson, D. H., Chewning, B., Bosworth, K., & Day, P. (1985). Interactive computer programs as public information campaigns for hard-to-reach populations: The BARN Project Example. *Journal of Communication, 37*(2), 8–28.

Hawkins, R. P., & Pingree, S. (1989). Adolescents, parents and new communication technologies. In H. Bertram, R. Borrmann-Muller, S. Hubner-Funk, & A. Weidacher (Eds.), *Adolescents and families* (pp. 459–472). Munich: Deutsches Jugendinstitut.

McDonald, M., & Blum, H. (1992). *Health in the information age: The emergence of health oriented telecommunication applications.* Berkeley, CA: Environmental Science and Policy Institute.

Moos R. H., & Schaeffer, J. (1984). The crisis of physical illness: An overview and conceptual approach. In R. H. Moos (Ed.), *Coping with physical illness II: New perspectives.* New York: Plenum.

Pingree, S., Hawkins, R. P., Gustafson, D. H., Boberg, E. W., Bricker, E., Wise, M., Behre, H., & Hsu, E. (1996). Will the disadvantaged ride the information highway: Hopeful answers from a computer-based health crisis system. *Journal of Broadcasting and Electronic Media, 40*(3), 331–353.

Rappoport, A. (1953). What is information? *ETC: The Journal of General Semantics, 10*(4), 5–12.

Schramm, W. (1955). Information theory and mass communication. *Journalism Quarterly, 32,* 131–146.

Sainfort, F. C., Gustafson, D. H., Bosworth, K., & Hawkins, R. P. (1990). Decision support systems' effectiveness: Conceptual framework and empirical evaluation. *Organizational Behavior and Human Decision Processes, 45,* 232–252.

Salmon, C. (1989). Campaigns for social "improvement:" An overview of values, rationales, and impacts. In C. Salmon (Ed.), *Information campaigns: Balancing social values and social change* (pp. 19–53). Newbury Park, CA: Sage.

Stewart, A. L., Hays, R. D., Ware, J. E. (1989). The MOS short-form general health survey: Reliability and validity in a patient population. *Medical Care, 26*(7), 724–732.

Strecher, V. J., McEvoy-DeVellis, B., Becker, M. H., & Rosenstock, I. M. (1986). The role of self-efficacy in achieving health behavior change. *Health Education Quarterly, 13,* 73–91

6

Interactive Video Games for Health Promotion: Effects on Knowledge, Self-Efficacy, Social Support, and Health

Debra A. Lieberman
Media Research Consultant

Video games offer unique advantages for delivering health promotion messages to children and adolescents. Although many of today's popular video games portray graphic violence and negative social stereotypes, video games can be designed to avoid these problems and instead promote more desirable skills and learning. Video games oriented to health promotion, for example, can help players learn about prevention and self-care and improve their health-related skills and behaviors. The video game format lends itself well to health promotion because games offer unlimited chances for repetition and rehearsal, and messages can be individualized to each player based on performance in the game. To engage young people in health-related behaviors while they play, video games can present appealing role-model characters, provide scenarios that involve making health decisions and carrying out self-care skills, and depict realistic consequences in response to players' decisions and actions.

Video games can reach young people who are unmotivated to learn about health, because playing video games—including health-oriented video games—is a very attractive leisure activity. In one study, children who were given a health promotion video game to keep at home spent an average of 36 hours playing it during a 6-month period (Lieberman & Brown, 1996). If an educational game is entertaining and offers a challenge that is neither too easy nor too difficult, many youngsters will play repeatedly over the course of several weeks or months; and many of them will remain interested in playing the game until they are able to complete it easily (Lieberman, 1995; Lieberman & Brown, 1995; Malone & Lepper, 1987). In the course of achieving this proficiency, young people necessarily learn the content embedded in the game. Our research presumes that experiential learning, integrated into the process of playing and mastering a video game, becomes deeply embedded in the person as well (see Brown et al., 1997).

This chapter considers how multimedia instructional design principles integrated into video games can contribute to health promotion targeted to young people. It points to relevant theory and research evidence in this area, and then considers how to design health promotion learning activities when games are accessible on networks online.

The discussion of design principles is based on research findings involving four titles from the Health Hero video game series, one with an anti-smoking message, two about diabetes self-management, and one about asthma self-management. These video games were produced by Raya Systems, a company in Mountain View, CA, that specializes in multimedia for health and learning. Health educators and multimedia instructional designers from Raya Systems created the video games with artists and software engineers from game developers WaveQuest and Sculptured Software, and with medical experts from Stanford University Medical Center, Stanford University Center for Research in Disease Prevention, Kaiser Permanente Northern California Region, and California Pacific Medical Center.

To support research and development, Raya Systems obtained funding from agencies, including the National Institute of Diabetes and Digestive and Kidney Disease, the National Institute of Allergy and Infectious Diseases, the U.S. Public Health Service Agency for Health Care Policy and Research, and the State of California Tobacco-Related Disease Research Program. Medical organizations, such as the American Academy of Pediatrics, the Juvenile Diabetes Foundation, and the Asthma and Allergy Foundation of America, endorsed and publicized the video games. Corporations, such as Novo Nordisk, Boehringer Mannheim Diagnostics, and Astra, purchased bulk orders of Health Hero video games and distributed them to health care providers, libraries, community organizations, and asthma summer camps.

FOUR HEALTH HERO VIDEO GAMES

Health Hero video games present memorable interactive simulations in which players must make prevention or self-management decisions for characters while also maneuvering those characters within a typical action-adventure game. Throughout the game, players observe and play the part of role-model hero characters, rehearse health-related skills in a simulated environment, receive both immediate and cumulative performance feedback, and make decisions that bring about realistic health consequences for the characters.

Rex Ronan—Experimental Surgeon is a Health Hero video game that illustrates the detrimental effects of smoking by taking players on a tour of a smoker's body. The game is designed to solidify players' negative attitudes about tobacco use and strengthen their intentions not to start smoking.

In the video game, players assume the identity of Dr. Rex Ronan, a brilliant surgeon who sets out to save a smoker's life. Shrinking to near-microscopic size and traveling to nine parts of the body, Dr. Ronan uses his laser scalpel to blast away phlegm, tar, plaque, and precancerous cells, all caused by smoking. Players must maneuver Dr. Ronan in and around realistic body organs, identify the unwanted deposits, and ultimately conquer the patient's nicotine addiction while also trying to avoid evil microbots who are out to thwart the life-saving mission.

In addition to presenting a graphic view of the physical effects of smoking, the video game conveys facts about its dangers. Players who correctly answer questions about the risks of tobacco use have a better chance of completing the game. The game is targeted to young people ages 10 to 16, the age group most likely to start smoking.

For diabetes, *Captain Novolin* is a self-management video game that demonstrates the relationship between food, insulin, and blood glucose control. The player assumes the role of Captain Novolin, a superhero who has diabetes and is involved in an action-packed battle with junk-food invaders while he is on his way to rescue plane crash victims stranded on Mount Wayupthar. To win the game, players must make wise choices about Captain Novolin's food, insulin doses, and blood glucose monitoring. As in many video games, there are dangers to avoid; Captain Novolin must dodge an onslaught of sugary donuts, ice cream, candy, and soda, all dangerous to young people who have diabetes. Players have opportunities to choose more appropriate foods instead.

Captain Novolin's blood glucose level fluctuates according to the insulin and food choices the player has made for him. Cumulative feedback on Captain Novolin's recent blood glucose measurements is provided throughout the game. If blood glucose goes too high or too low, Captain Novolin cannot vanquish his junk-food foes and save the crash victims in time. This provides the player with valuable experience in monitoring and managing diet and its consequences. Insulin-dependent diabetes often begins in childhood, and this game is targeted to ages 8 to 14.

In *Packy & Marlon*, two elephant pals who have diabetes head out for a diabetes summer camp. When they arrive, they find that rats and mice have invaded the camp and have scattered the food and diabetes supplies everywhere. Taking the role of Packy in a single-player game—and adding Marlon when a second player joins in—players help the heroes avoid enemy rodents, select healthy foods according to a recommended menu based on food exchanges, take insulin appropriately, and monitor blood glucose seven times in each of four simulated days. Players can access a chart that lists all the foods available in the game, highlight one food at a time, and see a display of the food exchanges in one serving. Players also answer questions dealing with diabetes self-management and how to handle problematic social situations. They can progress in the game only after they have selected the correct answer.

To find out how well they are managing Packy or Marlon's diabetes, players can retrieve a screen displaying a diabetes logbook, which shows a record of their character's recent blood glucose levels, the times of day the player gave the character insulin, how much insulin was given, and the food exchanges the player selected for each meal and snack. Players win points when they have managed Packy or Marlon's diabetes well. Good diabetes management also keeps the heroes in top form so players will have their best chance to complete and win the game. *Packy & Marlon* is targeted to ages 8 to 14.

Bronkie the Bronchiasaurus provides role-playing simulations to teach young people, ages 8 to 14, about asthma self-care. Players must help Bronkie and Trakie, two dinosaurs who have asthma, keep their peak flow (breath strength) at optimal levels while they avoid asthma triggers—and save their planet from dust and destruction.

The story begins in prehistoric times, when meteors struck the planet of San Saurian and left it engulfed in clouds of dust. To survive the disaster, the inhabitants developed a wind machine to clear dust from the air. Then, an evil Tyrannosaurus Rex stole the machine's core, locked it in his office, and scattered the remaining machine parts all over the city, lakes, jungle, sky, canyons, and caves. Bronkie, a boy bronchiasaurus, and Trakie, a girl tracheratops, set out to find the missing machine parts, outsmart the dinosaur thugs who are protecting the machine parts, and put the wind machine together before the dust returns.

Players of *Bronkie the Bronchiasaurus* can choose to play either Bronkie or Trakie in a single-player game, or opt for a two-player game. To manage Bronkie and Trakie's asthma, players must help the heroes take their daily asthma medicine with an inhaler and spacer at the beginning of each of the 18 game levels, and watch a detailed animation of Bronkie or Trakie going through each step of the process. During the game, players must maneuver the heroes so they avoid bumping into asthma triggers, such as furry animals, pollen, chemical drips, smoke, dust, and Sneezers who periodically sneeze cold viruses on anyone who gets too close. Peak flow decreases any time a hero comes in contact with a trigger. When peak flow goes down, the character coughs more frequently, the screen gets increasingly dim, and the player must find and administer emergency or sick-day medication appropriately. These survival sequences are not only typical of other popular video games, they resemble the daily experiences and concerns of many asthmatic children and adolescents.

Two questions about asthma self-management appear in each game level, and players have access to background information to help them answer the questions and proceed in the adventure. To monitor their progress, players can look at their asthma record, a screen showing a cumulative log of medications taken and peak flow measurements recorded throughout the game. Players earn more points if they make the right decisions and manage their character's asthma optimally, and they have the best chance of winning the game when their character's asthma is under control.

THEORY-BASED DESIGN
AND RELATED RESEARCH FINDINGS

We designed the Health Hero video game series to encourage healthy behaviors involving either prevention of health problems or self-management of ongoing conditions. Smoking is one of the first prevention topics in the Health Hero series because in early adolescence many young people hold ardent antismoking attitudes, but then they begin to regard smoking as an acceptable or even desirable thing to do. This age, when many youngsters begin smoking and encouraging one another to do so, also constitutes the age group that spends a great deal of time playing video games. Diabetes and asthma were chosen as self-management topics because children with these conditions must make self-management choices throughout the day, and failure to make the right choices can lead to serious health problems. It is impossible for parents to monitor their children's activities closely all the time; children with diabetes or asthma must be constantly responsible for being aware of early warning signs of physical problems, knowing what to do in emergencies, using medication regularly and appropriately, and making wise decisions such as choosing the right food to eat in the case of diabetes or avoiding environmental triggers that can bring on an asthma episode.

To help young people improve their health behaviors and outcomes, Health Hero video games focus on enhancing the young person's attention to and active processing of health promotion messages, motivation to learn about health, knowledge about prevention and self-management, perceived self-efficacy regarding healthy behaviors, and communication about health with others who can provide social support. Research in the fields of education, public health, and human–computer interaction, and in other cognitive and social sciences, has demonstrated that these intervening factors are important components in the success of health promotion interventions. Table 6.1 outlines some of the goals and related design features

TABLE 6.1
Goals and Related Design Features in Health Hero Video Games

Goals Are Addressed By	→	Design Features
Attention and active processing		
To reduce psychological distance; increase attention to the content; and make the content seem personally relevant to young people	→	Present content on the popular video game medium in a format young people perceive to be targeted to them.
To boost players' self-esteem; increase attention; optimize credibility; and increase the likelihood that young people will emulate the character's behaviors	→	Use attractive, competent role-model characters who have the same health condition as the target user group and are about 2 to 3 years older.
To increase attention, involvement, learning, and retention	→	Provide cognitive challenges, compelling characters and relationships, experiential learning, user control over the action, and individualized feedback.
Motivation		
To make games motivating, engaging, and appealing	→	Present clear, intriguing, and challenging goals, and provide continuous updates on progress toward the goals; provide individualized interaction and feedback.
To enhance enjoyment and individualize the learning experience	→	Allow game players to customize the content according to preferences and to match their own health status (e.g., player can select the frequency and dose of a diabetic character's daily insulin).
Knowledge		
To teach explicit content	→	Use direct instruction; include game strategies that require the player to learn information in order to succeed in the game; use very graphic and memorable illustrations such as disgusting tar, plaque, and debris shown in a smoker's body.
To teach skills	→	Present animated demonstrations such as how to use an inhaler for asthma medication; provide opportunities to rehearse skills and solve problems in simulations that show realistic outcomes based on the player's actions.

in Health Hero video games, and following it is discussion of the five intervening factors.

Attention and Active Processing

Children typically become very attentive to and involved in video games and other educational activities that are experiential and offer new or unusual experiences (Austin, 1995; Larson, 1991; Lepper & Gurtner, 1989; Lieberman & Linn, 1991). Other video game features that encourage attention and active processing include cognitive challenges, user control over the action, compelling characters and relationships, interactivity with immediate and individualized responses to the player, and the excitement

TABLE 6.1
(continued)

Goals Are Addressed By	→	Design Features
To ensure that players will retain the information and skills they have learned in the game	→	Repeat information and animated demonstrations for review when players give a wrong answer; make the game difficult enough that players will repeat game levels dozens of times and therefore will be exposed repeatedly to the same content.
To correct mistakes and improve performance	→	Provide constructive feedback about the player's actions and choices, and offer remediation as needed.
To provide background information on demand	→	Enable easy access to dynamic databases such as a food chart showing the food exchanges in a serving of each food that players may select in the game.
To provide a cumulative record of performance in the game; increase the player's understanding; and encourage the use of personal logbooks	→	Use on-screen, automatically updated logbooks that record, for instance, medications the character has taken and blood glucose or peak flow measurements attained in each game level.
Perceived self-efficacy		
To increase player's perceived self-efficacy for prevention and self-care	→	Create opportunities for players to rehearse new skills and to apply new knowledge in the game until they are successful.
To help players feel more confident and willing to discuss their health concerns with peers, parents, and caregivers	→	Present issues and questions that players must address in the game, thereby allowing them to rehearse the answers while playing alone or to discuss the answers when others are present.
Communication and social support		
To encourage social interaction that can increase peer tutoring, learning, and retention	→	Offer a two-player option in the game.
To provide a springboard for discussion about prevention or self-management	→	Create an appealing game that young people will want to talk about and will be proud to play.

of reacting quickly to events (Ambron & Hooper, 1990; Clements, 1987; McNeil & Nelson, 1991).

Simulations in a video game are scenarios in which players can make decisions, take action, observe the consequences, change course, and practice until they succeed. Players engage in problem solving in the simulated environment and see realistic outcomes. The challenges and goals posed in simulations often engage players in deep cognitive processing and learning-for-its-own-sake, which can be pleasurable and motivating experiences (Kozma, 1991; Malone & Lepper, 1987).

The *Bronkie the Bronchiasaurus* video game was designed to be a simulation that challenges players to manage the main character's asthma. Players must learn about asthma to make decisions appropriate to the environment. A variety of asthma triggers appear in the path of action: cigarette smoke, pollen, furry animals, dust, and Sneezers who emit cold viruses. Players learn to recognize and avoid the asthma triggers, read a peak flow meter, and take emergency or sick-day medication if they bump into several triggers that cause peak flow to go down. The simulation is realistic and the game environment changes predictably whenever the character bumps into a trigger (peak flow ebbs a certain amount, depending on the trigger) or takes asthma medication (peak flow increases a certain amount, depending on the medication).

A general premise underlying all Health Hero video games is that the style of the content and the selection of the medium itself can reduce the psychological distance between the message and the targeted audience. For example, adolescents may not be interested in a text-based health pamphlet that does not seem personally relevant to them. The same content in a video game, which they typically perceive as a young people's medium, will demonstrate that the message is targeted directly to their group. When psychological distance is reduced this way, audience members' attention and comprehension improve (Parrott, 1995). Educational video games, in addition to appearing on a young people's medium, can reduce psychological distance by including characters, actions, relationships, humor, settings, artwork, and music created intentionally to appeal to the target age group.

Motivation to Learn About Health

Video games provide clear and explicit goals, and this is motivating and attractive to youngsters (Malone & Lepper, 1987). Health goals can be integrated into a video game so that players frequently make health decisions and receive immediate and cumulative performance feedback about those decisions. Being motivated to win the game, players are inherently motivated to learn whatever is needed to help them perform well and win.

Children say they like to learn with video games. In three studies of Health Hero video games, interviews were conducted individually with young people who were asked first whether they would prefer to see a book, video game, or video tape about the health topic. In each study, more than 90% of the participants said they would prefer to see a video game instead of the other media suggested (see Lieberman, 1995; Lieberman & Brown, 1995). Young people's actions reflect their words when it comes to media preferences. Research with Health Hero games shows that most children who have opportunities to play the games do so extensively and enthusiastically during leisure time at home, as outpatients in clinic waiting rooms, or as hospital inpatients (Lieberman, 1995).

Young people see many advantages of video games over other media for learning. In a study, children ages 6 to 12 from diverse ethnic and economic backgrounds were asked if they saw any difference between learning with books, video games, or video tapes (Lieberman, 1993). Each interview was conducted individually, yet the children consistently pointed particularly to involvement and interactivity as features they like in video games:

"Video games put the person in the environment." (Boy, age 12)

"A video game makes you watch carefully." (Boy, age 11)

"A video game is more fun. There's action, not just words." (Girl, age 11)

"A book can explain, but you have the experience in a video game." (Girl, age 8)

"With a video game you get to practice more." (Girl, age 7)

"A video game tells you if you're wrong, so you can learn." (Girl, age 6)

"You can play in a video game. With books and videos you are only watching." (Boy, age 6)

These children prefer active participation in learning and they like receiving individualized feedback. With video games, they know exactly what they are in for, and they like it.

Knowledge About Prevention and Self-Care

Health promotion interventions have a greater chance of succeeding if they offer recipients of the message an opportunity to practice what they have learned (Bandura, 1986; Maibach & Cotton, 1995). The video game medium is equally suited to provide both instruction and opportunities for practice: Games can motivate players by requiring them to apply new knowledge to problem situations within the game environment. As with other multimedia, video game players are especially likely to develop new knowledge and skills when the software enables them to proceed at their own ability level and pace, requires that they revisit the content until they understand it thoroughly, and provides individualized and constructive performance feedback (Kulik & Kulik, 1991; Lieberman, 1992; McNeil & Nelson, 1991; Osman & Hannafin, 1993). We built these features into Health Hero video games to encourage and support learning.

Video games based on game levels are inherently repetitious. A player must start at the beginning of a game and progress through each level until the entire, lengthy sequence is completed. When the game is lost, players must start again from the beginning and, with a compelling game, they usually do this dozens or even hundreds of times. Health Hero games repeatedly deliver simple, fundamental health messages by capitalizing on the repetitiousness of game play. In the *Rex Ronan* game, for example, players constantly see harmful effects of smoking on organs in the human body. Players of the diabetes self-management game, *Packy & Marlon*, have 24 chances to choose a diet well-balanced in food exchanges and to receive feedback about their progress. In the asthma video game, a detailed and accurate animation of Bronkie and Trakie taking their medicine with an inhaler appears at the beginning of 18 game levels.

Social learning theory (Bandura, 1982, 1986) posits that behaviors can be learned both from observing the actions of others and from observing the outcomes and consequences of those actions. Young people learn how to behave when they observe others who are role models. Role models can be real people or characters portrayed in the media in both realistic video and cartoon renditions. When role models perform a behavior and then experience beneficial outcomes and consequences, those watching will in turn be more inclined to enact that behavior themselves. They will be more attentive to, and will be more likely to emulate, the behavior of role models who are most similar to them. We have applied these principles of social learning in Health Hero video games. For example, the main characters who deal with diabetes and asthma self-management are young and have the same health condition as the target group. For children and adolescents, role model characters should be about 2 or 3 years older than the target group. An age gap is desirable because young people are eager to learn how they are going to behave when they get a little older, so they look for slightly older role models to observe.

The target age ranges for Health Hero video games span about 7 years (ages 8 to 14 or 10 to 16). However, research on these games consistently finds that children as young as 5 and as old as 18 play the games competently and enthusiastically, and at all ages they learn a considerable amount about the health topic while playing the games (Lieberman, 1995).

Field tests and formative evaluations have found significant gains in children's knowledge when they played Health Hero video games about safety (Lieberman, 1993), smoking and its effects (Lieberman & Brown, 1995), and asthma (Lieberman, 1995). For example, in a study with 50 asthmatic children, ages 7 to 15, knowledge about asthma increased significantly after they had played *Bronkie the Bronchiasaurus* for 45 minutes. This knowledge was retained 1 month later in a delayed posttest interview, even though participants did not have access to the game during that month (Lieberman, 1995).

Another study found that the video game *Rex Ronan—Experimental Surgeon* elicited strong feelings in children about harmful effects of tobacco use. During game play, many players commented spontaneously about the harmfulness of smoking and what it does to the body. After playing, many expressed strong interest in learning more about the effects of smoking (Lieberman & Brown, 1995).

Perceived Self-Efficacy for Health Behaviors

Video games can be designed to enhance perceived self-efficacy, which refers to beliefs in one's own ability to carry out a task or successfully take a course of action needed to meet the demands of a situation (Bandura, 1982, 1986; Maibach & Cotton, 1995). Self-efficacy is predictive of future behavior. Confidence in one's own ability to perform a behavior links knowing what to do with actually doing it. So, when self-efficacy for a particular health behavior is high, the individual's health knowledge is more likely to translate into positive health behavior (Bandura, 1990; Maibach, Flora, & Nass, 1991). Improvement in health self-efficacy perceptions is an important goal for health promotion interventions such as video games, which provide an environment where young people can experiment, perhaps fail, but ultimately succeed. As they become more successful, they are likely to perceive themselves as being more efficacious in the tasks and skills they are rehearsing (Clements, 1987; Niemiec & Walberg, 1987).

We designed *Bronkie the Bronchiasaurus* to enhance self-efficacy by providing a way for players to rehearse asthma self-management until they become successful. In doing so, they gain confidence that they can manage asthma efficaciously in their own daily life. In a study of 50 asthmatic children and adolescents aged 7 to 15, participants felt more capable of taking care of their asthma and of communicating about asthma after they played the game (Lieberman, 1995). Self-efficacy increased for avoiding asthma triggers, for knowing what to do if peak flow goes down, for using emergency medications appropriately, and for talking with friends about asthma. These gains in self-efficacy were still significant in a delayed posttest 1 month later.

Self-esteem is another self-concept addressed in Health Hero video games. The self-management games for diabetes and asthma are designed to strengthen the self-esteem of children who have chronic conditions by featuring appealing and successful characters who are a lot like them. Studies have shown that young people are highly attentive to characters who are similar to them and they feel validated to see those characters represented in the media (Johnston & Ettema, 1986; McDermott & Greenberg, 1985).

Communication and Social Support

Without social support, a person is much less likely to stay healthy or to cope effectively with health problems that do occur (Peterson & Stunkard, 1989). Video games can encourage social interaction with peers and family members, as many children enjoy playing with others in competitive and collaborative games (Malone & Lepper, 1987; Salomon & Gardner, 1986). A health-oriented video game can help young people talk with others about their own concerns and can increase the likelihood that they will seek support and advice (Lieberman & Brown, 1995).

The *Captain Novolin* video game was evaluated in a study with 23 parent–child pairs, in which each child was aged 8 to 14 and had diabetes. In individual interviews, the majority of the children said that it would be easy to play the game with their friends. Several commented, without prompting from the interviewers, that the game would help them explain diabetes to their friends. All parents approved of the game and appreciated its potential to help them discuss diabetes with their children (Brown, 1993).

In another study, 50 asthmatic children, aged 7 to 15, showed significant gains in their frequency of communication about asthma with their friends and their parents after playing *Bronkie the Bronchiasaurus* for 45 minutes (Lieberman, 1995). The study compared the number of friends with whom they spoke about asthma (child's list of friends) during the week prior to playing the game and, in a delayed posttest 1 month later, the week prior to the posttest interview. The study also compared the number of conversations about asthma they had with their parents (parent's estimate) the month before they played the game (pretest) and the month after they played it (delayed posttest).

Bronkie the Bronchiasaurus offers a two-player option that allows children to interact with each other while they interact with the game. They can talk about asthma in the game context, and this is expected to help them discuss asthma more easily. Even when children are not playing, discussion about the game can be a springboard for conversation about asthma with friends, family members, and caregivers.

Many parents and clinicians enjoy playing Health Hero video games with children. The adults typically say they like to play the video games as a way to break the ice and to enable young people to discuss their health problems more freely (Lieberman, 1995). They also enjoy letting young players have a chance to be the expert, which is almost always the case with video games.

Table 6.2 presents a summary of research findings about Health Hero video games.

The findings outlined in Table 6.2 are especially noteworthy in the aggregate. This accumulation of research evidence demonstrates that Health Hero video games are successfully meeting the stated goals. When carefully designed with players' needs and interests in mind, video games for health promotion are interventions that work.

ONLINE ENHANCEMENTS

Online channels, such as the Internet, broadband networks, and some forms of interactive television, are becoming more widely accessible to a larger home audience, and multimedia learning activities and video games are among their program options. The following are suggestions for enhancing health promotion activities and games in networked multimedia environments.

Video games can be gateways to information searching. Although stand-alone video games on portable storage media, such as diskettes or cartridges, can hold a limited amount of information, players have access to extensive information resources when they play a game online. A health education video game online could require players to search for information in a medical encyclopedia or bibliographic database and then apply the information in the game. This could, for example, involve solving a mystery, answering a question, identifying a health problem after observing a few symptoms, or finding key information to complete a task in a simulation. Players can also use multimedia publishing tools to create their own stories or presentations based on elements in the game.

As health information becomes more available online (see Lieberman, 1992; Robinson & Walters, 1986), many health care professionals and

TABLE 6.2
Research Findings About Health Hero Video Games

Hypothesis: Young People Who Play a Health Hero Video Game:	Study Participants; Research Design	Research Finding
Attention and active processing		
Enjoy playing the game.	77 children, age 10–12; posttest	Participants played an antismoking video game for 45 minutes and reported very high enjoyment.
Spend many hours playing when they have the opportunity.	26 children with diabetes, age 8–15; posttest	Participants had a diabetes video game at home for 6 months and played it for 36 hours, on average.
Spend many hours watching others play when it is not their turn.	22 hospitalized children, 6 who had asthma, age 4–18; case studies	When an asthma video game was available in a hospital pediatric ward, children played it often and for hours at a time. When it was not their turn to play, many children spent hours watching other children play.
Motivation		
Are highly motivated to play.	187 children, age 5–15; in four posttest studies	Participants played a health video game for about an hour, and were almost always highly involved and very resistant to ending the session.
Knowledge		
Strengthen their attitudes about appropriate prevention behavior.	20 children, age 6–12; pretest-posttest	Participants played an antismoking video game for 1 hour. After playing, they expressed significantly stronger intentions not to smoke.
Say they prefer to learn about health with video games, not print or video.	30 children, age 5–12; pretest interviews	More than 90% of participants said they would prefer to learn with video games, which they consider more involving than books or video.
Learn about health.	187 children, age 5–15; in four pretest-posttest studies	Participants always scored significantly higher on health knowledge tests, on average, after they played health video games. Topics were safety, smoking, diabetes, and asthma
Perceived self-efficacy		
Improve in self-efficacy for managing their chronic condition.	50 children with asthma, age 7–15; pretest, posttest, delayed posttest	Participants improved in self-efficacy for asthma self-management after they played an asthma video game. Improvement was equally high 1 month later.
Improve in self-efficacy for communicating with others about their chronic condition.	50 children with asthma, age 7–15; pretest, posttest, delayed posttest	Participants improved in self-efficacy for talking with friends about asthma after they played an asthma video. Improvement was equally high 1 month later.
Communication and social support		
Talk more often with friends about their chronic condition.	26 children with diabetes, age 8–15; pretest-posttest	Participants named the friends with whom they talked about diabetes during the previous week. The average number of friends increased significantly after they had a diabetes video game at home for 3 months.
Talk more often with family members about their chronic condition.	50 children with asthma, age 7–15; pretest, posttest, delayed posttest	Parents reported the number of times their child spoke about asthma with them during the past month before the child played the asthma video game 1 time, and in a delayed posttest 1 month later. The average number significantly increased the month after their child played the game.

consumers want to learn how to use these resources. Game formats could offer opportunities for practice, with scenarios in which the player must make hypothetical decisions about elective surgery or first aid, or decide whether a medical problem is serious enough to merit an appointment with a doctor. The game could offer hints and pointers to help the player find essential information and learn how to navigate in each online medical resource.

Video games can be gateways to other people, as playing video games is often a social occasion. Young people like to participate in two-player versions of games and look over the shoulders of other game players (Lieberman, 1995). Online, it is possible to play a health promotion video game with another player who is in a distant location. Players can communicate online to find new friends who want to play a certain game and then, during game play, they can see the current status of the two-player game on their own screen. Between games, they can communicate online with friends who share similar game skills and preferences or who have health problems in common. The diabetes video game *Packy & Marlon*, for example, could enable two children with diabetes to play the game and, in the process, share game tips, diabetes self-management tips, and social support.

Health support groups can hold real-time chats online and participants can post messages in newsgroups or bulletin boards, which provide asynchronous communication. Health care professionals can participate in patient support groups to address questions and concerns and point to other sources of information. An online connection can be an optional communication link for a group that also meets face-to-face, or it can be the sole source of contact with the group. Anticipated developments in sound and video transmission on the Internet, and lower-cost video conferencing options, will bring more social presence to online chats, which now are usually text-based.

Functional learning environments can be created online. Functional learning involves engaging in personally relevant activities that serve a useful purpose. People are highly motivated to learn when information will help them carry out a task that interests them. There is a trend in U.S. education to offer more opportunities for functional learning in a meaningful context, and online communication offers new ways to provide this type of learning environment (Repman, 1993; Riel, 1985, 1989). It could involve writing an online health newsletter that others will read, teaching others about health topics, providing support, or working on collaborative studies and sharing data. Online networks are powerful learning resources because they combine information access and connections to other people.

Although individualization is possible on stand-alone multimedia applications, there are greater opportunities to individualize health education multimedia content when it is online. In Health Hero video games, players can select a language (English, Spanish, or, for *Packy & Marlon*, French); adjust the character's therapy (insulin dose and frequency in the two diabetes games); and play a single-player or a two-player version of the game (in *Packy & Marlon* and in *Bronkie the Bronchiasaurus*). Many other forms of individualization are also possible online because storage capacity is more extensive there. Individualization could include adjustments in reading level or in difficulty of game play, for example. Questions about the health topic, posed to the player throughout the game, could be chosen dynamically so that if a player gets one set of questions correct, a new set of questions could be substituted to make the game more challenging and varied. Game developers can also change the game more easily online because there is one central point of transmission. They can add new installments of a health education video game so that it grows and changes over time. This could encourage players to come back periodically to see what is new.

Online environments also offer more expanded opportunities for data collection for both users and developers. For example, educational video games can be designed to show players detailed analyses of their performance over time. Players can compare their performance with other players if scores are posted online. Game developers can benefit by collecting aggregate usage data online, to identify a particular group's interests, needs, and abilities, and then use that information to improve future games.

SUMMARY

Video games and other interactive media offer new opportunities to give children and adolescents effective health promotion messages. Research on Health Hero video games demonstrates that video games can improve several intervening factors that lead to better health outcomes for young people: attention to and active processing of health promotion messages, motivation to learn about health, knowledge about prevention and self-management, perceived self-efficacy for healthy behaviors, and communication about health with others who can provide social support. Online environments offer new ways to make learning about health more engaging, social, personally relevant, and individualized. Health educators and health communicators should consider integrating interactive media, such as stand-alone and online video games, into health interventions targeted to young people.

REFERENCES

Ambron, S., & Hooper, K. (Eds.). (1990). *Learning with interactive multimedia.* Redmond, WA: Microsoft Press.

Austin, E. W. (1995). Reaching young audiences: Developmental considerations in designing health messages. In E. W. Maibach, & R. L. Parrott (Eds.), *Designing health messages: Approaches from communication theory and public health practice* (pp. 114–144). Thousand Oaks, CA: Sage.

Bandura, A. (1982). Self-efficacy mechanism in human agency. *American Psychologist, 37,* 122–147.

Bandura, A. (1986). *Social foundations of thought and action: A social cognitive theory.* Englewood Cliffs, NJ: Prentice-Hall.

Bandura, A. (1990). Self-efficacy mechanism in physiological activation and health-promoting behavior. In J. Madden, IV (Ed.), *Neurobiology of learning, emotion and affect* (pp. 103–119). New York: Raven.

Brown, S. J. (1993). *Field test of Captain Novolin with 23 children.* Unpublished manuscript. Mountain View, CA: Raya Systems

Brown, S. J., Lieberman, D. A., Gemeny, B. A., Fan, Y. C., Wilson, D. M., & Pasta, D. J. (1997). Educational video game for juvenile diabetes self-care: Results of a controlled trial. *Medical Informatics, 21*(4).

Clements, D. (1987). Computers and young children: A review of research. *Young Children, 43*(1), 34–43.

Johnston, J., & Ettema, J. S. (1986). Using television to best advantage: Research for prosocial television. In D. Zillmann & J. Bryant (Eds.), *Perspectives on media effects* (pp. 143–164). Hillsdale, NJ: Lawrence Erlbaum Associates.

Kozma, R. B. (1991). Learning with media. *Review of Educational Research, 61*(2), 179–211.

Kulik, C. C., & Kulik, J. A. (1991). Effectiveness of computer-based instruction: An updated analysis. *Computers in Human Behavior, 7,* 75–94.

Larson, M. S. (1991). Health-related messages embedded in prime-time television entertainment. *Health Communication, 3,* 175–184.

Lepper, M. R., & Gurtner, J. (1989). Children and computers: Approaching the twenty-first century. *American Psychologist, 44*(2), 170–178.

Lieberman, D. A. (1995). *Three studies of an asthma education video game.* Report to the National Institute of Allergy and Infectious Diseases, National Institutes of Health, Bethesda, MD.

Lieberman, D. A. (1993). *Formative evaluation of a video game to teach children about safety rules and injury prevention.* Report to the National Institute of Child Health and Human Development, National Institutes of Health, Bethesda, MD.

Lieberman, D. A. (1992). The computer's potential role in health education. *Health Communication, 4*(3), 211–225.

Lieberman, D. A., & Brown, S. J. (1996). *Video game to promote diabetes self-management: Results of a six-month outcome study.* Report to the National Institute of Diabetes and Digestive and Kidney Diseases, National Institutes of Health, Bethesda, MD.

Lieberman, D. A., & Brown, S. J. (1995). Designing interactive video games for children's health education. In K. Morgan, R. M. Satava, H. B. Sieburg, R. Matthews, & J. P. Christensen (Eds.), *Interactive technology and the new paradigm for healthcare (pp. 201–210).* Amsterdam: IOS Press.

Lieberman, D. A., & Linn, M. (1991). Learning to learn revisited: Computers and the development of self-directed learning skills. *Journal of Research on Computing in Education, 23*(3), 373–394.

Maibach, E. W., & Cotton, D. (1995). Moving people to behavior change. In E. W. Maibach & R. L. Parrott (Eds.), *Designing health messages: Approaches from communication theory and public health practice* (pp. 41–64). Thousand Oaks, CA: Sage.

Maibach, E., Flora, J. A., & Nass, C. (1991). Changes in self-efficacy and health behavior in response to a minimal contact community health campaign. *Health Communication, 3*(2), 1–15.

Malone, T. W., & Lepper, M. R. (1987). Making learning fun: A taxonomy of intrinsic motivations for learning. In R. E. Snow & M. J. Farr (Eds.), *Aptitude, learning and instruction III, Conative and affective process analyses* (pp. 223–253). Hillsdale, NJ: Lawrence Erlbaum Associates.

McDermott, S., & Greenberg, B. (1985). Parents, peers, and television as determinants of black children's esteem. In R. Bostrom (Ed.), *Communication yearbook* (Vol. 8). Beverly Hills, CA: Sage.

McNeil, B. J., & Nelson, K. R. (1991). Meta-analysis of interactive video instruction: A 10 year review of achievement effects. *Journal of Computer-Based Instruction, 18*(1), 1–6.

Niemiec, R., & Walberg, H. J. (1987). Comparative effects of computer-assisted instruction: A synthesis of reviews. *Journal of Educational Computing Research, 3*(1), 19–35.

Osman, M. E., & Hannafin, M. J. (1993). Metacognition research and theory: Analysis and implications for instructional design. *Educational Technology Research and Development, 40*(2), 83–99.

Parrott, R. L. (1995). Motivation to attend to health messages: Presentation of content and linguistic considerations. In E. W. Maibach & R. L. Parrott (Eds.), *Designing health messages: Approaches from communication theory and public health practice* (pp. 2–23). Thousand Oaks, CA: Sage.

Peterson, C., & Stunkard, A. J. (1989). Personal control and health promotion. *Social Science and Medicine, 28*, 819–828.

Repman, J. (1993). Collaborative, computer-based learning: Cognitive and affective outcomes. *Journal of Educational Computing Research, 9*(2), 149–163.

Riel, M. (1989). The impact of computers in classrooms. *Journal of Research on Computing in Education, 22*, 180–190.

Riel, M. (1985). The Computer Chronicles Newswire: A functional learning environment for acquiring literacy skills. *Journal of Educational Computing Research, 1*, 317–337.

Robinson, T. N., & Walters, P. A. (1986). Health-Net: An interactive computer network for campus health promotion. *Journal of American College Health, 34*(6), 284–285.

Salomon, G., & Gardner, H. (1986). The computer as educator: Lessons from television research. *Educational Researcher, 15*, 13–19.

7

Information Environments for Breast Cancer Education

Richard L. Street, Jr.
Texas A&M University and Health Science Center

Timothy Manning
Texas A&M University Health Science Center

Contributors to medical journals and the popular press are for the most part optimistic, if not enthusiastic about the prospects of using interactive technology for health promotion (see, for example, Booker, 1996; Jelovsek & Adebonjo, 1993; Kahn, 1993). From the user's perspective, interactive computer programs can be fun, engaging, novel, and used in accordance with one's individual needs and interests. From the provider's perspective, interactive technology is a resource for developing cost-efficient and modifiable information systems for patient education and health promotion initiatives.

However, as noted in chapter 1, research comparing the effectiveness of various health promotion media indicates interactive technology is at times better and at times no more effective than less expensive, more traditional methods such as brochures and videotapes. The inconsistency of these findings is in part due to two limitations in the development and formative evaluation of computer programs designed for health and patient education. First, relatively few studies have compared and contrasted the effectiveness of various media for delivering health-related messages. Second, of the comparative studies that have been conducted, many fail to examine how characteristics of users (e.g., demographics, interest in the topic, emotional state), media (e.g., interactivity, multimodality), and message content (e.g., the topic, arguments, source of the message) collectively influence message processing and the subsequent effects of message processing on the user's health beliefs, attitudes, and behaviors.

In this chapter, we address these issues as they apply to information environments for breast cancer education. As mentioned in chapters 1 and 4, information environments refer to computer applications designed to provide users opportunities to actively explore information on a certain topic by creating a sense of presence within the media. Information environments typically consist of various databases presented through an array of media formats (e.g., motion picture, text, music, narration, graphics, computer animation) and an interactive interface that allows the user considerable control over the path taken within the environment.

THEORETICAL MODEL

The model presented in this chapter attempts to elaborate on certain features of the conceptual framework presented in chapter 1. In that model, we proposed that there are three stages of processes that ultimately influence the effectiveness of health promotion using interactive media: institutional, user, and technological variables affecting implementation and use; the user–media–message interaction and its effects on the user's knowledge, self-efficacy (e.g., the belief that one can enact specific health behaviors), motivation, and affective responses (e.g., comfort, reassurance, enjoyment); and other social, cultural, and economic factors affecting health behaviors and outcomes of interest. In this chapter, we examine the second and third stages as they relate to two different applications of breast cancer education.

Figure 7.1 presents a model of health promotion using information environments. We propose that the effectiveness of health messages designed to inform or persuade is contingent on the juxtaposition of two contexts of experience. One is the mediated information environment within which the message is presented. The other is the health behavior environment that consists of other social, cultural, and economic factors that also will influence health behaviors and outcomes of interest. The mediated information environment is the core component of the model and should be applicable to most uses of media for patient education and health promotion. The health behavior context will vary, however, depending on the particular outcome of interest. A major premise in the model is that the effectiveness of the health promotion or education effort will depend on the extent to which the educational outcomes (e.g., knowledge, motivation, problem-solving skills, self-efficacy) are achieved and are sufficiently powerful to elicit the targeted outcome in spite of (or in conjunction with) other factors within the health behavior environment.

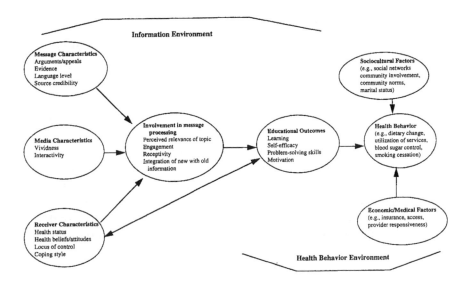

FIG. 7.1. A model of health promotion using information environments.

Consider, for example, two very different applications of breast cancer education (described in detail later). The goal of one program is to help patients newly diagnosed breast cancer patients become more involved in their consultations to decide local treatment (e.g., breast removal or breast conservation). The goal of the second application is to encourage breast cancer screening (e.g., performing a self-examination, having a mammogram) among women who do not routinely engage in this practice. With respect to the user–message–media interaction, both applications attempt to educate, enhance the user's sense of self-efficacy, and motivate the user to engage in a particular behavior. However, the targeted behaviors of interest—actively participating in deciding treatment versus scheduling a clinical breast exam or mammogram—emerge from very different health behavior environments. Patient involvement in deciding treatment will be affected by interpersonal and cultural variables such as how the physician interacts with the patient and the patient's attitudes about the doctor's and patient's roles in the encounter (Street, Voigt, Geyer, Manning, & Swanson, 1995). Whether a woman has a mammogram will depend on economic factors such as cost and availability and on personal factors such as knowledge and attitudes toward breast cancer screening (Vernon, Laville, & Jackson, 1990). To be effective, the educational intervention must be adapted and responsive to those exigencies uniquely related to the behavior of interest.

Mediated Information Environment

A premise embraced by the model assumes that educational messages are more effective when receivers are involved in message processing. As a construct, *involvement* has been variously defined as an internal state, as a property of a message, as a characteristic of the media, and as the perceived salience of the topic (Chaffe & Roser, 1986). For our purposes, we consider involvement to be a feature of message processing. Specifically, people are highly involved in message processing when they find the topic personally relevant, important, and interesting; are engaged by the message; closely attend to the arguments, evidence, and reasoning in the message; and actively integrate the information with their existing knowledge structures (Dede & Fontana, 1995; Petty & Cacioppo, 1986). When these conditions are not met, the message will likely have little effect on the receiver's current knowledge, attitudes, and behavior.

At least three sets of variables can influence involvement in message processing. First, there are individual differences in the extent to which people find a particular health issue personally relevant and important. These differences may predispose individuals to be more or less attentive to messages on the topic. For example, a person whose spouse smokes may closely examine information on the harms of secondhand smoke. However, a family of nonsmokers may consider the issue personally irrelevant, and a person who smokes may avoid the topic and thus the message. Some audience characteristics that may influence predispositions toward involvement in an issue are health status, lifestyle, sociodemographic variables, attitudes toward the topic, locus of control, and coping style (Miller, 1987; Steptoe & Appels, 1989; Strecher, DeVellis, Becker, & Rosenstock, 1986; Strickland, 1978).

Message characteristics also will have a powerful effect on the extent to which message recipients are involved in message processing and on whether the health messages accomplish their educational and behavioral objectives. Important message characteristics include the number and quality of arguments, nature of the evidence, organization of the message, redundancy of information, reading level, and word choice, to name a few (see Maibach & Parrott, 1995; Petty & Cacioppo, 1986; Stiff, 1986, for a review of these issues). Because we are examining the potential benefits of interactive media compared to other media for health promotion, we do not provide a detailed review of how message characteristics may affect information processing. However, it is important to emphasize that message content must be designed and articulated so that it makes a cogent case (e.g., with clarity, evidence, and reasoning) for learning about a health issue or engaging in a particular health behavior, is within the receiver's literacy level, and uses message sources (e.g., similar others, experts) that are respected and perceived as credible (Maibach & Parrot, 1995). Unfortunately, these criteria often are not met in health promotion and patient education materials (Davis, Crouch, Wills, Miller, & Abdehou, 1990; Reid, Kardash, Robinson, & Scholes, 1994; Webber, 1990).

Finally, media characteristics also can influence involvement in processing health messages. Rimal and Flora (chapter 2, this volume) provide an in-depth discussion of various dimensions of media and their potential relationships to learning, self-efficacy, and behavior. For the purposes of this chapter, however, we focus on two general features of media, *vividness,* and *interactivity* (Biocca, 1992; Steuer, 1992). *Vividness* represents the richness of the mediated environment in terms of the number of media components used (e.g., video, narration, text, music, animation). An information environment that utilizes more components can create a more vivid presentation in terms of the number of senses engaged in message processing (e.g., sight, sound, touch), the sensory quality of the stimulus (e.g., color, movement, resolution, loudness), and the extent to which message content is accentuated (e.g., using narration and video to demonstrate a behavior). *Interactivity*, on the other hand, refers to the extent to which the mediated environment is responsive to and controlled by the user's actions. Responsiveness represents the degree to which the media allows responses back to the user and the form, content, and nature of this response (e.g., corrective feedback, questions, answers, a move in a game; Rafaeli, 1988). User control allows the individual to modify the form and content of the mediated environment by determining what information is selected, what information is repeated, and the order in which these selections are made (Steuer, 1992).

Health Behavior Environment

The features of this portion of the model are situation-specific in that they represent the array of personal, social, cultural, and economic factors that may influence the outcome of interest (see Fig. 7.1). For example, the variables that will determine whether a patient becomes involved in medical decision making will differ to some extent from those factors affecting whether an individual utilizes cancer screening services. As Skinner and Krueter mentioned in chapter 3, researchers and program evaluators should select the theory or model that provides the best insight into how a particular health promotion intervention can be applied so that the user's experience within the information environment more effectively contributes to health behaviors and outcomes of interest.

INFORMATION ENVIRONMENTS
FOR BREAST CANCER EDUCATION

We apply this model in a formative evaluation of two interactive multimedia programs for breast cancer education. Options for Breast Cancer Treatment was designed for women newly diagnosed with early breast cancer. Options for Better Breast Health focuses on breast cancer screening and the importance of early detection. For several reasons, we believed that interactive multimedia programs would more effectively accomplish our educational and health promotion objectives than would more traditional media such as brochures and videotapes.

For example, a number of studies indicate that media that are more sensory rich (e.g., videotape, television, videodisc) and interactive (interactive multimedia, computer games) facilitate learning and (sometimes) behavior change more than do less vivid and interactive media (e.g., pamphlets, videotapes; Gagliano, 1988; Jelovsek & Adebonojo, 1993; see Schaffer & Hannafin, 1986; chapters 1 & 6, this volume) presumably because they enhance attention to the information, active learning, and individualization of content (Dede & Fontana, 1995; Kahn, 1993; Skinner, Siegfried, Kegler, & Strecher, 1993). Also, more difficult informational content may be made more amenable to comprehension if several media components are integrated (e.g., narration, text, computer animation) to convey and enhance ideas and themes in the message.

However, as depicted in our conceptual model (Fig. 7.1), other factors extraneous to the intervention also may affect outcomes of interest (e.g., social relationships, cultural attitudes, economic resources). Thus, health promotion interventions must be designed and applied to either accommodate or change these exigencies in order to enhance the likelihood of a successful outcome. The two studies described next were designed to test the feasibility of this model.

Options for Breast Cancer Treatment

This program describes options for treating early breast cancer. The information environment was constructed using a combination of text, narration, music, graphics, motion pictures, and computer animation. The program is divided into four sections: Introduction, Understanding the Problem, Treatment Options, and Experiences of Other Women. The Introduction describes how to use the program and presents brief audiovisual clips of eight women sharing their initial reactions to the cancer diagnosis. Understanding the Problem briefly overviews the organs of the breast (e.g., milk ducts, mammary glands, areola), what breast cancer is and where it may form, and the various stages of breast cancer.

Treatment Options includes information on two options for treatment: mastectomy and lumpectomy with irradiation. On several occasions, text and narration state that the two treatments are comparable in their effectiveness and likelihood for cure. This section also displays a side-by-side comparison of the advantages of each treatment (e.g., lumpectomy preserves the breast; mastectomy is the quicker form of treatment). Experiences of Other Women includes audiovisual clips of eight women (four who chose lumpectomy and four who chose mastectomy) sharing their experiences about treatment, their decisions, and the recovery process. At several points throughout the program, the text and audio narration encourage the patient to ask questions, express concerns, and offer opinions when they visit with doctors.

Using a touchscreen monitor, patients can explore the various topics in whatever order they choose. One of the few constraints is within the Treatment Options section. Once the patient chooses to review information on one form of treatment, she can not exit the section until she views information about the other treatment option. Of course, a patient can quit the program at any time. It takes about 35 to 45 minutes to run through the entire course of the program.

Theoretical Rationale

Although the program can be used for general education purposes, it was specifically designed as a preconsultation intervention for use after the patient has been informed of the diagnosis but before she visits with physicians to decide treatment. The educational goals of the program are to help patients learn about their treatment options, foster a positive outlook for their future, and motivate them to be active participants in their upcoming consultations. Working from our theoretical model, we believed that a multimedia program not only was appropriate for these objectives, it also would be superior to brochures, the most common method of patient education about breast cancer treatment.

First, although people newly diagnosed with cancer are certainly interested in their options for treatment, their emotional states (e.g., anxiety, fear, denial, shock) might interfere with the ability to read and concentrate on information presented in an assortment of pamphlets. A multimedia program, on the other hand, presents narration, text, music, and an array of visual images that could engage the patient and help focus attention to the message. Second, many breast cancer patients express frustration with the process by which treatment decisions are made (Hughes, 1993). Some patients say they received too little information, others report being overwhelmed by all the information provided, and still others are unsure of the patient's role in the decision-making process (Cawley, Kostic, & Cappello, 1990; Siminoff, Fetting, & Abeloff, 1989). A multimedia program could use multiple media components to enhance comprehension, and an interactive interface would allow patients to pick, choose, and review topics of interest. Finally, a multimedia presentation can present video clips of healthy breast cancer survivors describing their own experiences about treatment and recovery. This, in turn, could comfort and reassure the patient and perhaps foster a positive outlook for the future (Gustafson et al., 1993).

The behavioral outcomes of interest are to increase patient participation in the consultation (e.g., asking questions, offering opinions, expressing concerns) and involvement in deciding treatment. We hypothesized that a breast cancer patient would more actively discuss treatment issues when she had some knowledge about the topic and believed in the legitimacy of her participation in these discussions. We also thought that the patient's active exploration of the information environment (e.g., selecting topics, repeating topics, choosing her own course through the program) also might carryover to facilitate more active involvement in the consultation.

However, to be effective, the intervention must take into account other processes within the health behavior context. In this case, two other factors also could have a strong influence on patient participation. First, some patients (e.g., older, less educated) may be less inclined to take an active role in deciding treatment than are other patients (e.g., younger, more educated) (Cassileth, Zupkis, Sutton-Smith, & March, 1980; Street, 1992). Second, patients generally become more involved in their consultations when they perceive their physicians as allowing and encouraging patient participation (e.g., by avoiding interruptions, asking for the patient's opinion, attentive listening; Cox, 1989; Street, 1992). To address these issues, the multimedia program included several messages (in text and narration) stating that doctors were interested in the patient's opinions and concerns, that the choice for treatment largely depends on the patient's preferences, and that the patient should discuss these issues with the doctor.

The Study

Because the complete details of this study are reported elsewhere (Street et al., 1995), we provide only a brief summary here.

Procedures.   Sixty women with stage I or II breast cancer were randomly assigned to one of two preconsultation education conditions—brochure or multimedia computer program. The brochure contained the same medical information as the multimedia program and also encouraged women to openly discuss their concerns and opinions with the doctors. Prior to the education, patients completed measures assessing knowledge about breast cancer treatment and optimism for the future (Scheier & Carver, 1985). After receiving the education, patients completed the measures again and then visited with a medical oncologist, radiation oncologist, and surgeon. During the visit with the surgeon, a decision was made for local treatment. Patient involvement measures included the patient's perceptions of control over decision making (adapted from England & Evans, 1992), perceived involvement in the consultation (adapted from Lerman et al., 1990), and the frequency with which patients asked questions, offered opinions, and expressed concerns (coded from audiorecordings of the consultations).

Results.   Whereas patients using the multimedia program tended to score higher ($p = .07$) on the knowledge test (83% correct answers) than did patients reading the brochure (76% correct), the strongest effect on this measure was simply providing the preconsultation education. In other words, patients were significantly more knowledgeable about breast cancer treatment after receiving the education regardless of method (78% correct) than they were prior to the intervention (59% correct). Second, optimism was not affected by the educational interventions, as patients who were more optimistic at the outset of the study tended to be more optimistic after their consultations. However, optimism was moderately correlated with knowledge, $r = .34, p < .01$. Finally, patient involvement was not influenced by the method of preconsultation education but, as expected, was related to personal factors and to patients' perceptions of the physicians' behavior. Specifically, patients who were more active participants in these consultations were younger, more educated, and perceived their physicians as encouraging patient involvement.

Options for Better Breast Health

Our second program in the Options series is Options for Better Breast Health, an interactive multimedia program that describes various methods of detecting breast cancer and advocates routine screening. The program consists of text, graphics, music, narration, still pictures, computer animation, and motion pictures. This program covers six general topics—risk factors, cancerous and noncancerous breast lumps, clinical breast examination, self-examination, mammography, and biopsy. The best detection strategy is presented as a combination of breast self-examination, clinical examination, and mammography. With a mouse or touchpad, the user controls the time spent with each topic, the order in which the topics are presented, and whether to repeat topics. Most users take between 15 to 25 minutes to use the program.

Theoretical Rationale

Breast cancer screening varies greatly among women 40 years of age and older (Lerman, Rimer, Trock, Balshem, & Engstrom, 1990). Prochaska and DiClemente's (1986) theory of behavior change offers a useful theoretical framework for understanding the health behavior context of breast cancer screening. The model posits that individuals differ in their readiness to adopt health behaviors and that this variability may be represented along a continuum from precontemplation (not considering the behavior) to contemplation (considering the behavior) to action (have previously but not currently engaged in the behavior) to maintenance (actively engaged in the behavior).

Women who routinely screen for breast cancer consider this a matter of personal importance, are knowledgeable of various methods of early detection (e.g., mammogram, self-examination), talk to their doctors about these concerns, believe cancer is a disease they can control, and actively utilize screening services (Rimer, Trock, Engstrom, Lerman, & King, 1991). Conversely, women in the precontemplation stage tend to be unaware of the importance of early detection, think they are not at risk, are uncertain (and perhaps anxious) about screening procedures and their value, or perceive little control over the disease (Rakowski, Dube, Marcus, Prochaska, Velicer, & Abrams, 1992).

Women not up to date with screening guidelines are typically in the precontemplation or contemplation stage of behavior adoption (Rakowski et al., 1992; chapter 3, this volume). Thus, an important goal for breast cancer education is to facilitate the individual's "readiness to adopt" the intended behavior (i.e., precontemplation to contemplation) and provide the means and encouragement to engage in the activity (contemplation to action). We believed that, compared to videotapes and brochures, a multimedia program would more effectively accomplish these objectives.

For one thing, the interactive capabilities of the medium allows users to explore topics of concern at their own pace, thereby individualizing the instruction. Individualization of content is particularly important given that women differ in their reasons for not practicing breast cancer screening. For example, some women may lack the knowledge; others may be anxious about the procedures (Rimer et al., 1991). In addition, a program created with multiple media components could present the information in a dynamic, interesting, and engaging manner. If the presentational medium facilitates involvement in message processing, then the content of the message may have a better chance of increasing knowledge and motivation, reducing anxiety, and enhancing a sense of self-efficacy, all of which should increase the message recipient's intention to screen for breast cancer. If the intervention also provided an opportunity to act on these behavioral intentions (e.g., schedule a mammogram; pay a reduced fee; talk to the doctor about breast cancer screening), it would be even more effective (NCI Breast Cancer Screening Consortium, 1990).

To date, we have only tested Options for Better Breast Health using a sample of female college students. Not only was this group convenient, but we assumed that their level of involvement with breast cancer screening issues would vary and, on average, be rather low. Thus, the sample allowed us to examine whether an interactive multimedia program would engage a low-involved audience to a greater degree than would more traditional media such as brochures and videotape. Two research questions were examined: Do different types of media (brochure vs. videotape vs. multimedia program) affect the degree to which message recipients gain knowledge about breast cancer screening and report involvement in the topic? Do message recipients evaluate breast cancer education differently depending on their level of involvement in the topic and the modality of education?

The Study

Methods. A total of 77 undergraduate women participated in this research. The participants were primarily White with an average age of 23 (range = 19–41). The research was conducted over a 3-week period in a communication laboratory equipped with computers, VCRs, and private rooms for viewing or reading. Research participants first completed pre-education measures (knowledge about breast cancer screening, involvement in the topic), received one of three modalities of education (pamphlet, videotape, multimedia program), and then once again completed the knowledge and involvement measures. The research participants also evaluated the educational materials on separate scales.

Options for Better Breast Health served as the interactive multimedia presentation. For the videotape condition, a run-through of Options was downloaded onto a VHS tape. Thus, the videotape presented in a linear fashion one path through the program. For the text-only condition, the content document used to develop Options was rewritten so that it read like a brochure on breast cancer screening. The knowledge measure consisted of 16 multiple-choice questions assessing knowledge about breast cancer, noncancerous breast lumps, risk factors, and screening procedures. Involvement in the topic was assessed with six items asking participants to report the degree to which they found breast cancer to be a topic of personal concern, importance, interest, and a topic about which one should learn more and discuss with friends and relatives. Respondents evaluated the educational materials on five items assessing the degree to which the presentation was understandable, interesting, motivating, frustrating, and informative. (These measures are available from the authors on request.)

Data Analysis.   Two sets of data analysis were conducted. First, 2 (pre-education vs. post-education) × 3 (brochure, videotape, multimedia program) analyses of variance (ANOVAs) were performed on the knowledge and involvement measures in order to test whether these variables were influenced by the education per se (regardless of media) and by the type of media. Second, to test the possibility that preferences for a particular presentational media would depend on an individual's initial involvement in the topic, a median split was performed on the involvement scores to create two groups of participants who were either more or less involved in breast cancer screening. The means for the lower and higher involved groups were 19.4 and 24.5, respectively, on the 30-point scale. A 2 (high vs. low involved) × 3 (brochure, videotape, multimedia) ANOVA was then performed on the evaluative measures. It is important to note that the mean score for the low involvement group was still higher than the scale midpoint of 18. This indicates that, on average, these college women were concerned about and interested in breast cancer screening.

Results.   As in the other study, knowledge about breast cancer screening increased significantly after participants reviewed the educational materials ($F = 320.68$, $p < .0001$; $Ms = 48\%$ correct responses before the education and 85% correct responses after the education). Knowledge about breast cancer screening was not influenced by the modality of education, $F = 1.31$, NS. The results were similar for the involvement measure as these college women became more involved in the topic after the intervention, $F = 15.25$, $p < .0001$ $Ms = 21.8$ and 23.8 respectively for the pre- and post-education scores on the 30-point scale). The type of media $F = 2.12$, NS and the interaction between involvement and type of media, $F = .90$, NS, had no noticeable effect on involvement in the topic.

The second set of analyses produced a more complex pattern of results. First, there were main effects for method of presentation on two measures. Specifically, the multimedia program was rated significantly, $F = 8.95$, $p < .01$, less boring ($M = 2.02$ on a 5-point scale) than were the brochure ($M = 2.81$) and videotape ($M = 3.24$) but significantly, $F = 10.63$, $p < .01$ more frustrating ($M = 3.21$), than the other two presentations ($Ms = 1.63$ and 2.02 for the brochure and videotape). Second, there were significant interaction effects between degree of involvement and method of presentation on the extent to which the audience perceived the message as understandable, $F = 4.24$, $p < .05$, and motivating, $F = 3.32$, $p < .05$. Specifically, the low-involved group found the brochure ($M = 4.9$) and videotape ($M = 4.9$) easier to understand than they did the multimedia program ($M = 3.7$), and they perceived both the multimedia program ($M = 2.9$) and brochure ($M = 3.4$) as less motivating than they did the videotape presentation ($M = 4.2$). Conversely, the high-involved group found the presentations easy to understand regardless of media ($M = 4.9$, 4.9, and 4.7 for the brochure, videotape, and multimedia program respectively) and were highly motivated by the message, particularly when using the multimedia program ($Ms = 4.1$, 4.0, and 4.2 for the brochure, videotape, and multimedia program respectively).

CONCLUSIONS

The results of the formative evaluations of these programs offer several implications for our model of health promotion using information environments and for the development of health education materials.

Implications for Theory and Research

Educational Outcomes.   At the outset, we proposed that different media environments would influence the extent to which health promotion and educational messages facilitated learning, positive affect, motivation, and self-efficacy. The results of these investigations found little evidence of the superiority of one type of media relative to another. Rather, and perhaps not surprisingly, informational content (a message characteristic) had the strongest effect on knowledge acquisition and, in the case of breast cancer screening, self-reported involvement in the topic. Message content improved these outcomes regardless of whether it was delivered via text, a linear video presentation, or an interactive multimedia presentation.

However, as observed in the second investigation, an individual's a priori involvement in the topic to some extent influenced media preferences. Specifically, college women who initially were less concerned about breast cancer and less interested in screening found the videotape presentation more understandable and motivating than they did the multimedia program. Conversely, women who considered the topic of personal import found all the presentations to be motivating, easy to understand, interesting, and informative. In addition, more involved receivers spent about 5 minutes longer with the multimedia program ($M = 23$ min) than did the less involved users ($M = 18$ min).

These findings suggest that the two dimensions of media, sensory vividness and interactivity, may have different effects on media preferences depending on the audience's level of involvement. For example, someone with little interest in a particular health issue may have a passive or uncaring orientation to information on the topic. This person may perceive the interactive features of a multimedia program as requiring too much effort to use or as a means of avoiding information that is of little interest. However, such a person might become engaged by an audiovisual presentation because the sensory richness of the message (e.g., voice, music, color, movement, text, graphics) can arouse (and perhaps entertain) the passive or disinterested perceiver to the point of stimulating some curiosity about the topic. On the other hand, a person who is concerned about a certain health issue will actively examine information on the topic regardless of the presentational medium. Furthermore, such an individual may especially appreciate the interactive interface of a multimedia program because it allows exploration of the information according to his or her specific needs, interests, and concerns.

Third, choice of media may be a very important issue when message recipients are emotionally aroused. Individuals experiencing strong emotions (e.g., fear, anxiety) are easily distracted and often have difficulty concentrating on an educational message. However, a dynamic presentation that uses narration, computer animation, video, and an interactive interface, may engage the receiver and focus attention to the content of the message. Although the studies reported here were not designed to measure the influence of different media on emotions, we do have anecdotal evidence regarding the emotional impact of Options for Breast Cancer Treatment. The oncology nurse administering the project chose not to invite four patients to participate in the study because they were noticeably distraught and emotionally upset. However, these women did use the multimedia program before their appointments to discuss treatment. According to the nurse's report, these patients gradually became more calm and relaxed after they started the program and spent as much time exploring the information as did patients participating in the study. One wonders whether a brochure would have had the same effect.

Of course, we do not know why the program might have had a therapeutic effect. Was it the soft-spoken narration? Was it the sensory vividness of a multimedia presentation? Was it the audiovisual testimonials of breast cancer survivors? These findings suggest that, at least with respect to cognitive outcomes, a variety of media can achieve positive results when the message content is well-designed. However, how one copes with breast cancer and whether one engages in breast cancer screening also are influenced by such feelings as fear, anxiety, anger, and sadness. Selection of an appropriate educational media may well depend on the receiver's affective state, an issue currently needing much more research attention.

Behavioral Outcomes.   Only the first study reported here offered an opportunity to examine the effectiveness of different methods of patient education in facilitating behavior change. The degree to which breast cancer patients participated in their consultations to decide treatment was mostly related to the patients' personal characteristics (age and education) and to the doctor's efforts to facilitate the patient's involvement. There were no noticeable differences in patient involvement as a result of whether the patients read a brochure or used the multimedia program. Although other researchers have shown that preconsultation education can increase patient participation in consultations (Greenfield, Kaplan, & Ware, 1985; Rost, Flavin, Cole, & McGill, 1991), an important difference between those studies and our investigation is that we did not include a *no education* condition in the research design. In other words, all patients in our study received some form of preconsultation education that, in turn, may have contributed to their involvement in these consultations.

In an informal test of this notion, we surveyed 62 women at the same clinic who had received treatment for early breast cancer after data collection in the original study was completed. These women were asked to report on a nine-point scale how much control they believed they had over the decisions for treatment. Patients ($n = 17$) whose preconsultation education included both the multimedia program and brochure reported more control over decision making ($M = 7.1$) than did women who only received the brochure ($n = 25$, $M = 6.2$). In turn, the brochure-only group reported more decision control than did those patients ($n = 20$, $M = 5.7$) who did not receive preconsultation education. Although these differences only approached significance, $F = 2.44$, $p < .1$, there was a significant contrast ($p < .05$) between the multimedia-plus-brochure group and the group that did not receive preconsultation education. These findings provide some support for the notion that appropriate preconsultation education can facilitate patient participation in deciding treatment for early breast cancer and that multimedia technology may enhance this involvement.

Implications for Program Development

We do not repeat here what other chapters discuss regarding the development and implementation of interactive media for patient education and health promotion materials. However, given our findings, a couple of points should be highlighted. First, researchers and program developers must be very careful when designing a friendly interface (see also chapters 5 and 13). For example, some users of our program on breast cancer screening were initially confused by what the different menu icons represented although we who created the program thought the icons were very simple and self-explanatory. Some people, particularly those who are not familiar with computing technology, will need very clear instructions, even demonstrations, on how to use interactive programs and what the different icons, buttons, and menus represent. This should be a standard feature of programs with, of course, the option not to review instructions if the user chooses.

Second, one of the more interesting findings of the second study was that low-involved message recipients evaluated the videotaped presentation more favorably than they did the brochure or the multimedia program. For health topics in which audience involvement may be low, one might consider using a sensory rich but traditional medium, such as a linear audiovideo presentation, to introduce both the topic and the interactive features of the multimedia program. For example, a carefully crafted video presentation at the outset might intrigue the viewer and stimulate his or her interest in the topic. Then, the interactive features of the program could be offered for more active exploration of the information environment.

Finally, text that is well written (e.g., easy to read, clearly organized), narration that is understandable and pleasant, audiovisual sequences that show demonstrations, and graphical displays of information all potentially can contribute to improved cognitive outcomes such as knowledge acquisition. The vast majority of message recipients were engaged by and learned from these materials. However, it is important to recognize that these were recruited audiences in that they were asked to examine the materials, agreed to do so, and reviewed the materials in privacy. In effect, the situation was ideal for a close examination of the message. Although such environments may be obtainable in clinical settings, other forms of outreach may be less viewing-friendly. For example, shopping malls, supermarkets, and health fairs offer an array of displays and activities that may compete with or distract attention from any one particular health message. In such settings, the type of media may play an important role in engaging an audience. This may explain why videodiscs at a shopping mall or health fair often attract more attention than brochures left on a table.

REFERENCES

Biocca, F. (1992). Virtual reality technology: A tutorial. *Journal of Communication, 42*, 23–72.

Booker, E. (1996). What's your URL, Doc? *WebWeek, 2*, 47.

Cassileth, B. R., Zupkis, R. V., Sutton-Smith, K., & March, V. (1980). Information and participation preferences among cancer patients. *Annals of Internal Medicine, 92*, 832–836.

Cawley, M., Kostic, J., & Cappello, C. (1990). Informational and psychosocial needs of women choosing conservative surgery/primary radiation for early stage breast cancer. *Cancer Nursing, 13*, 90–94.

Chaffe, S. H., & Roser, C. (1986). Involvement and the consistency of knowledge, attitudes, and behavior. *Communication Research, 13*, 373–399.

Cox, A. (1989). Eliciting patients' feelings. In M. Stewart, & D. Roter (Eds.), *Communicating with medical patients* (pp. 99–106). Newbury Park, CA: Sage.

Davis T. C., Crouch M. A., Wills G., Miller S., & Abdehou D. M. (1990). The gap between patient reading comprehension and the readability of patient education materials. *Journal of Family Practice, 31*, 533–538.

Dede, C., & Fontana, L. (1995). Transforming health education via new media. In L. Harris (Ed.), *Health and the new media: Technologies transforming personal and public health* (pp. 163–184). Mahwah, NJ: Lawrence Erlbaum Associates.

England, S. L., & Evans, J. (1992). Patients' choices and perceptions after an invitation to participate in treatment decisions. *Social Science and Medicine, 34*, 1217–1225.

Gagliano M. (1988). A literature review of the efficacy of video in patient education. *Journal of Medical Education, 63*, 785–792.

Greenfield, S., Kaplan, S., & Ware, J. E., Jr. (1985). Expanding patient involvement in care. *Annals of Internal Medicine, 102*, 520–528.

Gustafson, D. H., Wise, M., McTavish, F., Taylor, J. O., Wolberg, W., Stewart, J., Smalley, R. V., & Bosworth, K. (1993). Development and pilot evaluation of a computer-based support system for women with breast cancer. *Journal of Psychosocial Oncology, 11*, 69–93.

Hughes, K. K. (1993). Decision-making by patients with breast cancer: The role of information in treatment selection. *Oncology Nursing Forum, 20*, 623–628.

Jelovsek, F. R., & Adebonojo, L. (1993). Learning principles applied to computer-assisted instruction. *M.D. Computing, 10*, 165–172.

Kahn, G. (1993). Computer-based patient education: A progress report. *M.D. Computing, 10*, 93–99.

Kozma, R. B. (1991). Learning with media. *Review of Educational Research, 61*, 179–211.

Lerman, C., Brody, D. S., Caputo, G. C., Smith, D. G., Lazaro, C.G., & Wolfson, H.G. (1990). Perceived involvement in care scale: Relationship to attitudes about illness and medical care. *Journal of General Internal Medicine, 5*, 29–33.

Lerman, C., Rimer, B., Trock, B., Balshem, A., & Engstrom, P. F. (1990). Factors associated with repeat adherence to breast cancer screening. *Preventive Medicine, 19*, 279–290.

Maibach, E., & Parrott, R. L. (Eds.). (1995). *Designing health messages.* Thousand Oaks, CA: Sage.

Miller, S. M. (1987). Monitoring and coping: Validity of a questionnaire assessing styles of information-seeking under stress. *Journal of Personality and Social Psychology, 52*, 345–53.

NCI Breast Cancer Screening Consortium. (1990). Screening mammography: A missed clinical opportunity. *Journal of the American Medical Association, 264*, 54–58.

Petty R. E., & Cacioppo J. T. (1986). *Communication and persuasion: Central and peripheral routes to attitude change.* New York: Springer-Verlag.

Prochaska, J. O., & DiClemente, C. C. (1986). *The transtheoretical approach: Crossing traditional boundaries of therapy.* Homewood, IL: Dow Jones-Irwin.

Rafaeli, S. (1988). Interactivity: From new media to communication. In R. P. Hawkins, J. M. Riemann, & S. Pingree (Eds.), *Advancing communication science: Merging mass and interpersonal processes* (pp. 110–134). Newbury Park, CA: Sage.

Rakowski, W., Dube, C. E., Marcus, B. H., Prochaska, J. O., Velicer, W. F., & Abrams, D. B. (1992). Assessing elements of women's decisions about mammography. *Health Psychology, 11*, 111–118.

Reid, J. C., Kardash, C. A., Robinson, R. D., & Scholes, R. (1994). Comprehension in patient literature: The importance of text and reader characteristics. *Health Communication, 6*, 327–335.

Rimer, B. K., Trock, B. T., Engstrom, P. F., Lerman, C., & King, E. (1991). Why do some women get regular mammograms? *American Journal of Preventive Medicine, 7*, 69–74.

Rost, K. M., Flavin, K. S., Cole, K., & McGill, J. B. (1991). Change in metabolic control and functional status after hospitalization: Impact of patient activation intervention in diabetic patients. *Diabetes Care, 14*, 881–889.

Schaffer, L. C., & Hannafin, M. J. (1986). The effects of progressive interactivity on learning from interactive video. *ECTJ, 34*, 89–96.

Scheier, M. F., & Carver, C. S. (1985). Optimism, coping, and health: Assessment and implications of generalized outcome expectancies. *Health Psychology, 4*, 219–247.

Siminoff, L. A., Fetting, J. H., & Abeloff, M. D. (1989). Doctor-patient communication about breast cancer adjuvant therapy. *Journal of Clinical Oncology, 7*, 1192–1200.

Skinner, C. S., Siegfried, J. C., Kegler, M. C., & Strecher, V. J. (1993). The potential of computers in patient education. *Patient Education and Counseling, 22*, 27–34.

Steptoe, A., & Appels, A. (Eds.). (1989). *Stress, personal control, and health.* Luxembourg: Wiley.

Steuer, J. (1992). Defining virtual reality: Dimensions determining telepresence. *Journal of Communication, 42*, 73–93.

Stiff, J. B. (1986). Cognitive processing of persuasive message cues. *Communication Monographs, 53*, 75–89.

Strecher, V. J., De Vellis, B. M., Becker, M. H., & Rosenstock, I. M. (1986). The role of self-efficacy in achieving health behavior change. *Health Education Quarterly, 13*, 73–91.

Street, R. L., Jr. (1992). Communicative styles and adaptations in physician-parent consultations. *Social Science and Medicine, 34*, 1155–1163.

Street, R. L., Jr., Voigt, B., Geyer, C., Manning, T., & Swanson, G. (1995). Increasing patient involvement in deciding treatment for early breast cancer. *Cancer, 76*, 2275–2285.

Strickland, B. R. (1978). Internal–external expectancies and health-related behaviors. *Journal of Clinical and Consulting Psychology, 46*, 1192–1211.

Vernon, S. W., Laville, E. A., & Jackson, G. L. (1990). Participation in breast cancer screening programs: A review. *Social Science and Medicine, 30*, 1107–1118.

Webber, G. C. (1990). Patient education: A review of the issues. *Medical Care, 28*, 1089–1103.

8

Patient-Specific Interfaces to Health and Decision-Making Information

Holly B. Jimison
Oregon Health Sciences University

Patients represent an extremely diverse set of users of computer systems for health information. They vary in their medical conditions and their values on quality of life, as well as in the experiential backgrounds they bring to the interaction with the computer. In the design of interactive technology for health information, it is important to determine how best to present health information to individual users. This chapter discusses a variety of methods currently used to tailor patient education materials using computers. Additionally, an approach for automating the creation of patient-specific materials is described in detail. This approach is based on computer decision modeling to measure the importance of variables in an individual patient's treatment decision is described in more detail.

PATIENT DIVERSITY AND INFORMATION NEEDS

Patient education researchers and practitioners have long recognized the value of tailoring the presentation of material to individual patients in face-to-face encounters (Hewson, 1993; Skinner, Siegfried, Kegler, & Strecher, 1993). Patient educators routinely incorporate their knowledge of differences in adapting their interactions with individual patients to accommodate for variations in cultural backgrounds and health beliefs, as well as variations in language ability and medical background. Additionally, patient differences in medical condition require that clinicians modify their explanation of treatment options to an individual's level of risk and potential benefit. However, on a routine basis, patients are more likely to receive only a small amount of educational information from their physicians, and perhaps a brochure to take home. The nature of a brochure as a vehicle for communicating patient education information requires that it be designed for the typical patient. The factors that physicians and health educators take into account with individual patients are difficult to replicate in a mass-produced brochure. The early computer-based approaches to patient education did little to improve upon brochures, but recent advances in authoring tools and computer hardware have led to the development of systems that have made progress toward emulating a sensitive and well-trained health educator with the ability to tailor the presentation of material to an individual's needs.

Cultural Aspects

Patients throughout the world, and even within the United States, come to the clinical encounter from widely disparate cultural backgrounds. The communication of information with the patient must account for health beliefs and quality of life priorities (Kleinman, Eisenberg, & Good, 1978). The framework for the Health Belief Model (Becker, 1974) characterizes the factors associated with preventive health action on the part of a patient. These factors include perceptions about the personal benefits and barriers relating to a health action as well as perceptions about the underlying personal health risk. A patient educator must be able to assess a patient's belief system and goals for change in order to be an effective advocate for the patient. For treatment planning, a clinician must also be able to accommodate varying patient backgrounds and values in coming to shared treatment goals. Education, negotiation, and motivation are important components of the provision of quality medical care. They directly influence adherence to treatment goals, which is in turn critical for improved health outcomes (Bird & Cohen-Cole, 1989; Lipkin, 1987). Delbanco (1992) and Waitzkin (1991) both emphasized the importance of incorporating patient preferences on quality-of-life issues into medical decision making and the construction of treatment goals.

The need to tailor the clinical encounter based on patient preferences and background extends to patient education. Whereas it is difficult to account for these differences in the print and graphics medium of commonly available patient education brochures, it is possible with computer tools to assess cultural background, beliefs, and preferences, and to address these differences with varying user interfaces. Thus far, the more typical approach with both existing computer tools and existing brochures is to target a specific audience with a specific product. In the future, we can expect that larger and more general systems will have user interfaces that adapt to a given user's personal and cultural preferences.

Language

A major barrier to effective patient–clinician communication, and even basic clinical care, is that in many communities, a large number of patients are not sufficiently fluent in English either to communicate or understand information about their basic medical needs. The general problem of requiring translators in many languages and dialects during an office visit is labor intensive and expensive. Many hospitals and clinics make use of telephone operator services and three-way headsets to aid communication for shorter periods of time. However, it becomes more difficult to include patient education using these methods. In some areas, printed brochures are available in Spanish or other sufficiently common languages. Yet, the inventory difficulties and expense associated with multiple versions of brochures can make a computer approach more attractive. Computer interfaces with easy access to similar information in multiple languages provide a useful solution to the translation problem. The basic data and program structures remain the same. Also, there are new software products to aid in translating English to other languages as a first pass to speed the new content development process.

Reading Level

Currently, the most commonly used format for patient education is the brochure or pamphlet, making heavy use of text. However, most studies on the literacy of the general population of the United States show that approximately 20% of the U.S. population is functionally illiterate (Hunter & Harman, 1985; Bureau of the Census, 1982) and that the reading comprehension of public clinic patients is about Grade 6.5 (Davis et al., 1991). Most studies measuring the understanding of patient education materials have found that only approximately 50% of the patients are unable to understand the written material (Vivian & Robertson, 1980). Patient comprehension is certainly a prerequisite to a patient's adherence to health care treatment goals, and the conventional techniques for patient education often fall short in this regard. Although guidelines for the development of patient education materials encourage using a reading level at or under the 6th or 8th grade level, most materials are at reading levels that far exceed that. Patient informed consent forms are often the most difficult to read, as they typically are written at the college level. Many of the written handouts for patients currently available could be improved by redesigning the content at a lower reading level; however, one of the advantages of computer-based approach to patient education is that through the use of video clips, graphics, and audio, it is possible to provide educational materials without requiring that the patient know how to read. Additionally, by assessing the user's approximate reading-level or desired level of complexity, it is possible for existing reading level metrics to limit the search of health databases to material at the appropriate level.

Level of Education and Medical Expertise

A variation of the readability problem has to do with a patient's ability to understand medical language. A highly educated patient may have very little medical background and not be familiar with much medical terminology. Conversely, a hospital or clinic employee with less formal education and reading ability could obtain a fairly high degree of understanding of medical terminology. Again, it is difficult to meet varying patient needs with a single brochure. Currently, most computer-based patient education products address this problem with built-in dictionaries with automated word look-up to assist patients in reading text on medical topics. Recent work on tailoring the searching of health information on the Internet includes developing user models that estimate a patient's level of education and medical experience (Pavel, Jimison, Anwar, Appleyard, & Sher, 1996).

POTENTIAL OF INTERACTIVE TECHNOLOGY TO ADDRESS DIVERSE PATIENT NEEDS

The technological demands for effective consumer health information and education are many and varied, especially when we include the need for video transmission and/or storage. The advances in computer hardware have been rapid, but the features of many of the existing computer-based patient education products have not yet made use of the newer capabilities. However, the greatest barrier to effective computer-assisted patient education is a software issue—the design of the user interface. This is also the area that holds the most promise for improving patient education. The concept of the user interface encompasses the aspects of how humans interact with the information available in a computer system (screen display, interactivity, etc.). The user interface of computer-assisted patient education enables interactivity and tailoring to individual patients (Tibbles, Lewis, Reisine, Rippey, & Donald, 1992; Vargo, 1991). Some of the basic potential advantages of a computer-based approach to patient education over traditional techniques are:

- Consistent content and delivery
- More easily available than a health educator
- Potentially cost effective (if compared to routine and consistent patient education by staff)
- Privacy of communication
- Active learning

In addition, the ability of computer systems to provide interactivity and tailor material to an individual user offers an additional important set of benefits:

- Self-paced instruction, more detail if desired
- Can repeat, review, and provide feedback on responses
- Tailor language, reading level, medical experience, and user preferences
- Tailoring based on health risk assessment
- Automated tailored record of patient understanding, knowledge and confidence
- Automated tailored record of patient education provided

Vargo (1991) and Tibbles et al. (1992) observed similar benefits but also noted concern about the lack of software availability as well as concern about system designs that preclude use by low-literacy patients and those fearful of computers. These issues are rapidly being addressed by developers of software for patient education. The Informed Patient Decisions Group's 1996 *Directory of Consumer Health Informatics* (Kieschnick, Adler, & Jimison, 1996) lists more than 600 software products. Many of these are designed for patients unfamiliar with computers and for those with low reading skills. The newer systems typically have user interfaces that are much easier to understand and that make greater use of video and graphics.

CURRENT CAPABILITIES OF COMPUTER-ASSISTED PATIENT EDUCATION TOOLS

Computer-assisted patient education tools are increasingly being designed with the capability to adapt to the needs of individual users. Such systems fall into four main categories:

- Tailored printouts of patient education materials

- Tailored searches of electronic databases of patient health information
- Interactive text-based tutorials
- Interactive multimedia systems

The health topics covered in such computer systems range the full spectrum, including:

- Wellness and preventive care (e.g., nutrition, exercise)
- Self-care, when to see a doctor
- Health risk assessment
- Symptoms, diseases, tests, and treatments
- Emergency and first aid
- Drug information
- Informed consent
- Advance directives

Systems for many of these medical domains inherently provide tailoring as a result of user interaction. For example, health risk appraisal systems use patient-specific information provided by the user to create a tailored summary of health risks. Most drug information systems allow the user to check for drug-drug interactions from a specific list of a patient's medications. Other systems will provide information on when to see a doctor based on a specific set of symptoms. Additionally, most computer-assisted patient education tools allow patients to browse, repeat, and review material as desired.

Some examples of systems that provide tailored printouts of patient education and health promotion come from the work of Strecher and his colleagues on tailored messages for smoking cessation (Strecher et al., 1994), mammography (Skinner, Strecher, & Hospers, 1994), and diet (Campbell et al., 1994). These are examples of systems with tailoring based on theoretical guidelines applying to patient education. These systems assess information about the patient's perceived benefits and barriers associated with a health behavior (Health Belief Model; Becker, 1974), the patient's readiness to change (Transtheoretical Model of Change; Prochaska & DiClemente, 1983), and the patient's perceived causes of success or failure in undertaking a new health behavior (Attribution Theory; Weiner, 1986). The contents of the resulting printed letters to patients then vary depending on the results of these assessments. For example, smokers in the precontemplative stage (not thinking about quitting yet) would only receive information about the perceived risks of smoking-related disease and the perceived benefits of quitting (health improvement, cost savings, etc.).

Information on potential barriers to quitting would not be addressed in this group. These researchers and developers have shown the tailored, printed material for patients to be more effective than traditional generic materials in promoting positive health behavior change.

To address the individual needs of users searching electronic databases of health information on the Internet, researchers at Oregon Graduate Institute and Oregon Health Sciences University have developed techniques for user modeling based on estimated reading level and medical expertise (Pavel et al., 1996). The user interface for searching health information is tailored to the abilities of an individual user either through optional direct assessment, where the user indicates the desired level of detail and medical terminology, or through inference of these levels based on the language of the user's queries. Statistical techniques are used to rank the appropriateness of the possible patient information based on a match of the language of the queries with the language of the relevant sections in databases of health information.

TAILORING BASED ON DECISION THEORETIC MODELS

A final approach to tailoring patient education materials that accounts for both medical differences among patients as well as differences in patients' values on the quality of life of health outcomes is described in more detail here. This method uses a decision-theoretic framework that provides a set of metrics for judging the importance of various personal and medical factors that relate to a patient's treatment decision. These measures of importance are then used to focus and tailor treatment and test information to a specific patient's needs. The theoretical grounding for this approach comes from decision theory, which embodies a methodology for quantifying the expected value to be gained from deciding to take one action over another based on probability theory, Bayes' theorem, and utility theory. It incorporates information about the likelihood of events based on taking a particular action, how those events in turn affect the likelihood of other outcomes, and the decision-maker's value associated with end outcomes. Given the correctness and completeness of the decision model, a rational decision maker would choose the alternative with the greatest expected value.

Jimison and colleagues used this framework and methodology in the development of the Angina Communication Tool, a system for educating patients about treatment and test alternatives for angina (Jimison, 1990;

Jimison, Fagan, Shachter, & Shortliffe, 1992). In this system, a decision-theoretic framework provides a set of metrics for judging the importance of various personal and medical factors that relate to a patient's treatment options for angina. The approach is based on a Bayesian network, such as the one shown in Fig. 8.1 for the Angina Communication Tool, to represent a decision model for the medical domain of interest. In this example, a comparison is made between bypass surgery (*Surgery*) versus medical management (*Meds*) for the treatment of angina (represented by the box in Fig. 8.1). *Overall Utility* (double outline in Fig. 8.1) is modeled as being influenced by several intervening variables that depend on both the treatment alternative and patient characteristics. The single outline nodes in this network contain information on prior and conditional probabilities, and the arrows between nodes show dependencies in probabilities.

Initially, the probabilities and values (utilities) represent a population average. A distribution, most often approximated by a normal distribution described with a mean and variance, is used to characterize how these values vary over a whole population. The entry of a sequence of information about the patient is used to tailor this population decision model to a more patient-specific model. With every new piece of information about the patient, the prior distributions of probabilities of possible events in the model and the prior distributions for likely patient utilities (patient preferences) for the possible health outcomes become more focused and better describe the probabilities and utilities for a particular patient. Using Monte Carlo simulation to obtain repeated sample probabilities and utilities from the distributions at each node, the expected utility for each treatment alternative is calculated and represented with a mean and confidence interval from multiple samples.

Figure 8.2 shows how the optimal treatment choice may change as the decision model becomes more tailored to the patient. The narrower distributions on the probabilities and utilities in the model also tighten the resulting distribution on the expected utilities for each choice of treatment. The patient assessment questions used to tailor the decision model are asked in an order most likely to separate the expected utilities of the treatment options. One of the benefits of this approach is that by monitoring the overlap of the confidence intervals on the expected utility distributions, the assessment process may be shortened to ask only the number of questions necessary to separate the alternatives and have a sufficiently robust decision.

Most often, such decision models are used for the sole purpose of rating one treatment alternative as preferable to another. However, the design of the Angina Communication Tool de-emphasized that aspect and instead

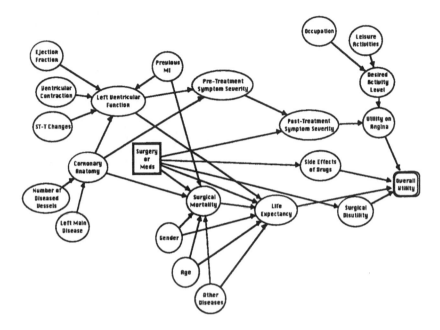

FIG. 8.1. Diagram of the Bayesian network representing the variables included in the decision model for the Angina Communication Tool.

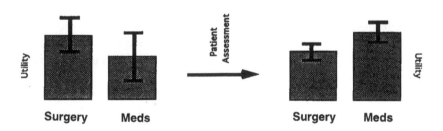

FIG. 8.2. Graphical representation of how the estimates for the expected utilities for surgical treatment versus medical management and the uncertainty around those estimates change with patient-specific information.

focused on using the decision theory metrics of expected value of information and sensitivity to tailor printed explanation handouts for a patient. Figure 8.3 shows a diagrammatic overview for a system to tailor patient handouts based on the output of a decision model. There are two generic

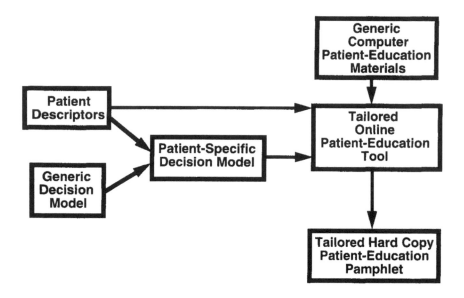

FIG. 8.3.   Overview of the system modules for tailoring patient education materials based on information from the refinement of a generic, population-based decision model to a patient-specific decision model.

modules in this system, the population decision module and the set of generic patient education templates. First, the generic population-based decision model is tailored using patient descriptors. Second, the output of the tailoring of the decision model (in the form of expected value-of-information and sensitivity of each of the variables) is in turn used to tailor the generic patient education materials.

Figures 8.4, 8.5, and 8.6 show the three types of patient education materials that may be accessed either through the system itself or in printed form. Figure 8.4 shows an example of the first class of patient education materials, background material that is left in generic form. A large section of the generic templates for patient education are appropriate and a necessary component for all patients. A second class of output is material that is tailored from direct patient input or material from the patient record. This may include native language (Spanish materials available in the Angina Communication Tool) or specific drug information as shown in Figure 8.5.

The final class of patient education materials is the set that is tailored through the output of the decision model. A tailored explanation of treatment options is automatically created to emphasize the following:

- Factors that have the largest influence in the treatment decision for that patient
- Factors that deviate from the norm of the general population
- Factors that still remain sensitive (small variation could change the decision).

Each of the criteria measures differences between the characteristics of the population model and the patient-specific model. The results of these measurements then determine the content of the printed explanation of treatment decisions and educational material. Figure 8.6 shows an example of instances where the text has been automatically tailored through decision-theory metrics (the underlines are for demonstration purposes). In this example, the risk of death from surgery was determined to be a sensitive variable (increased risk for coronary artery bypass surgery) and the primary influencing factor was advanced age, which was automatically derived by comparing the patient's age to the mean and standard deviation of the prior distribution in the population model for angina patients. The second paragraph contains a default preferred treatment from the decision model, which is designed to be easily edited by the physician. The listing of important

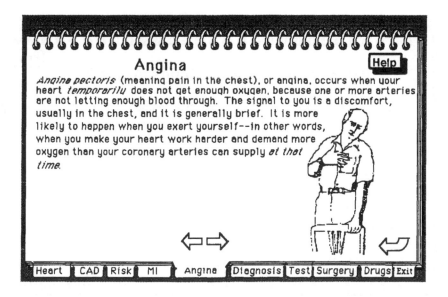

FIG. 8.4. Sample of general background information on angina for patients.

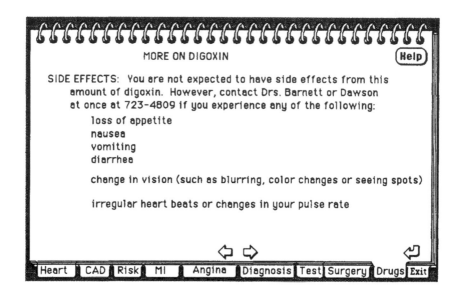

FIG. 8.5. Sample of tailoring based on direct information from the patient record.

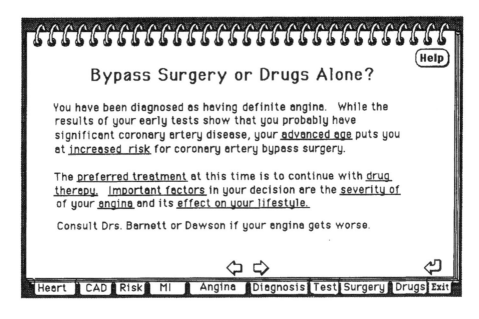

FIG. 8.6. Sample of tailored output based on output from the decision model.

factors to consider comes from an ordered ranking of the influence each variable in the model had on the decision. In this case, quality-of-life issues that affected the utility associated with the resulting health outcomes were most influential, and thus most important for the patient to discuss with his or her physician.

The purpose of this approach to the tailoring of patient education materials is to provide a focus on material that is important to an individual's treatment decision. This approach was tested with output from the Angina Communication Tool with cardiac patients making judgments on the quality and appropriateness of materials for hypothetical patients. The results showed a significant preference on the part of patients for tailored educational materials.

EFFECTIVENESS OF TRAINING
PATIENT EDUCATION MATERIALS

Several other studies confirm the effectiveness of tailoring patient education materials. In their review of the literature on evaluations of the effectiveness of patient education for chronic conditions, Mullen, Green, and Persinger (1985) found that tailoring or individualization was explicitly studied as part of the educational process in more than half of the evaluations that they reviewed ($n = 41$). These studies showed that individualization, along with feedback and reinforcement, were the strongest predictors of education quality. Additionally, many groups have looked at the overall benefits of using computers for patient education (Kulik & Kulik, 1991; Rippey et al., 1987; Sinclair, 1985; Tibbles et al., 1992; Vargo, 1991), and the main findings are that computer-based methods can be more effective and less costly when compared to traditional techniques. For the studies that specifically looked at tailoring material to individual patients through the use of computers (Jimison et al., 1992; Skinner et al., 1993; Skinner et al., 1994; Strecher et al., 1994), we see that patients prefer tailored information and actually have improved outcomes based on tailored educational interventions.

FUTURE DIRECTIONS IN TAILORED INTERFACES
TO HEALTH INFORMATION

Current trends in medical care include increased consumerism and efforts to control the escalating costs of medical care while maintaining the quality of medical care. Patients are more interested in participating in their medical care and treatment decisions, as well as taking an active role in searching for their own health information. At the same time, managed care organizations are interested in demand management to control utilization, including patient education efforts to encourage self-care and the appropriate utilization of medical services. All this points to a growing need for enhancing and coordinating efforts in patient education, health promotion, and the empowerment of patients. New communications technology, including networked computers and telemedicine systems, will play a large role in this effort. More medical care will be provided at a distance (in the home or in assisted living situations) with an increasing emphasis on long-term follow-up for both the medical care and patient self-care. The tailoring of the presentation of health information to the patient will take on increasing importance. Systems must be easy to use with information that is appropriate to the situation. There is a large role that computer tools for patient education can play in the changing medical environment, with prospects for both improving the quality of care and potentially assisting in the control of rising costs of care.

REFERENCES

Becker, M. H. (1974). The health belief model and personal health behavior. *Health Education Monograph, 2,* 324–473.

Bird, J., & Cohen-Cole, S. A. (1989). The three-function model of the medical interview. In M. S. Hale (Ed.), *Models of teaching Consultation-Liason Psychiatry,* (pp. 65–88). Basel, Switzerland: Karger.

Bureau of the Census. (1982). *English language proficiency study* (ELPS-4). Suitland, MD: Government Printing Office.

Campbell, M. K., DeVellis, B. M., Strecher, V. J., Ammerman, A., De Vellis, R. F., & Sandler, R. S. (1994). The impact of message tailoring on dietary behavior change for disease prevention in primary care settings. *American Journal of Public Health, 84*(5), 783–787.

Davis, T. C., Crouch, M. A., Long, S. W., Jackson, R. H., Bates, P., George, R. B., & Bairnsfather, L. E. (1991). Rapid assessment of literacy levels of adult primary care patients. *Family Medicine, 23*(6), 433–435.

Delbanco, T. L. (1992). Enriching the doctor-patient relationship by inviting the patient's perspective. *Annals of Internal Medicine, 116,* 414–418.

Hewson, M. G. (1993). Patient education through teaching for conceptual change. *Journal of General Internal Medicine, 8*(7), 393–398.

Hunter, C. S., & Harman, D. (1985). *Adult illiteracy in the United States: A report to the Ford Foundation.* New York: McGraw-Hill.

Jimison, H. (1990). Generating explanations of decision models based on an augmented representation of uncertainty. In R. D. Shachter, L. N. Kanal, T. S. Levitt, & J. F. Lemmer (Eds.), *Uncertainty in Artificial Intelligence* (Vol. 4, pp. 351–365). North Holland: Elsevier.

Jimison, H., Fagan, L. M., Shachter, R. D., & Shortliffe, E. H. (1992). Patient-specific explanation in models of chronic disease. *Artificial Intelligence in Medicine, 4*(3) 191–205.

Kieschnick, T., Adler, L. A., & Jimison, H. B. (1996). *1996 Health Informatics Directory.* Baltimore: Williams & Wilkins.

Kleinman, A., Eisenberg, L., & Good, B. (1978). Culture, illness, and care: Clinical lessons from anthropologic and cross-cultural research. *Annals of Internal Medicine, 88,* 251–258.

Kulik, C. I., & Kulik, J. (1991). Effectiveness of computer-based instruction: An update analysis. *Computers in Human Behavior, 7*(1), 75–94.

Lipkin, M. J. (1987). The medical interview and related skills. In W. T. Branch & M. J. Lipkin (Eds.), *Office practice of medicine* (pp. 1287–1306). Philadelphia: Saunders.

Mullen, P. D., Green, L. W., & Persinger, G. S. (1985). Clinical trials of patient education for chronic conditions: A comparative meta-analysis of intervention types. *Preventive Medicine, 14,* 753–781.

Pavel, M., Jimison, H. B., Anwar, J., Appleyard, R., & Sher, P. P. (1996). *User models for adaptive searching of patient health information on the internet* (Tech. Rep.). Portland: Oregon Graduate Institute.

Prochaska, J. O., & DiClemente, C. C. (1983). Stages and processes of self-change of smoking: Toward an integrative model of change. *Journal of Consulting and Clinical Psychology, 51,* 390–395.

Rippey, R. M., Bill, D., Abeles, M., Day, J., Downing, D. S., Pfeiffer, C. A., Thal, S. E., , & Wetsone, S. L. (1987). Computer-based patient education for older persons with osteoarthritis. *Arthritis and Rheumatism, 30*(8), 932–935.

Sinclair, V. (1985). A randomized controlled trial of a new approach to preoperative teaching and patient compliance. *International Journal of Nursing Studies, 22*(2), 105–114.

Skinner, C. S., Siegfried, J. C., Kegler, M. C., & Strecher, V. J. (1993). The potential of computers in patient education. *Patient Education and Counseling, 22*(1), 27–34.

Skinner, C. S., Strecher, V. J., & Hospers, H. (1994). Physicians' recommendations for mammography: Do tailored messages make a difference? *American Journal of Public Health, 84*(1), 43–49.

Strecher, V. J., Kreuter, M., Den Boer, D. J., Kobrin, S., Hospers, H. J., & Skinner, C. S. (1994). The effects of computer-tailored smoking cessation messages in family practice settings. *Journal of Family Practice, 39*(3), 262–270.

Tibbles, L., Lewis, C., Reisine, S., Rippey, R., & Donald, M. (1992). Computer assisted instruction for preoperative and postoperative patient education in joint replacement surgery. *Computers in Nursing, 10*(5), 208–212.

Vargo, G. (1991). Computer assisted patient education in the ambulatory care setting. *Computers in Nursing, 9*(5), 168–169.

Vivian, A. S., & Robertson, E. J. (1980). Readability of patient education materials. *Clinical Therapy, 3,* 129–136.

Waitzkin, H. (1991). *The politics of medical encounters: How patients and doctors deal with social problems.* New Haven, CT: Yale University Press.

Weiner, B. (1986). *An attributional theory of motivation and emotion.* New York: Springer-Verlag.

9

Health Promotion, Social Support, and Computer Networks

Patricia Flatley Brennan
University of Wisconsin–Madison

Sue V. Fink
University of Michigan

Health promotion is good for the health of individuals and for public health. Behaviors that are under individual control such as exercise, good nutrition, and not smoking promote health—that is, they both reduce the risk of disease and enhance the quality of life. Although these behaviors are controlled by the individual, the concerted efforts of professionals and the person's social support network may be required. These efforts are directed toward creating an awareness of the needs and opportunities for health promotion, engaging people in strategies to define and select behaviors congruent with their health promoting goals, and enabling people to integrate these behaviors into lifestyle patterns. Although the success of behavioral interventions has been demonstrated and the strengthening effects of social support documented, the abilities of computer technologies to facilitate the development of healthy lifestyles have not been fully explored.

ASPECTS OF HEALTH PROMOTION

Health promoting behaviors include behaviors that enhance health and well-being, reduce health risks, and prevent disease. Health promoting behaviors include regular exercise, diets adequate in nutritional components, and development of an intact self-concept. Risk-reduction behaviors include smoking cessation, wearing seat belts, avoiding foods likely to lead to dietary imbalance, and developing skills needed to control chronic illnesses. Health protection behaviors, such as immunizations and dental hygiene, are designed to avoid illness or prevent further decline of one's health state.

Some health promoting behaviors are enabling behaviors that, once incorporated into a life pattern, provide the foundation for healthy, satisfied living. They include exercise and positive mental thought. Other behaviors may be classified as avoiding behaviors; in these, the health promotion occurs not because the behavior is engaged in but because it is avoided. Both enabling and avoiding behaviors have social dimensions. For example, it is very difficult to avoid certain foods or eating patterns recognized as interfering with health when these foods and patterns are an important part of family and social relations. When an individual tries new behavior, family and friends may support the change and help to sustain it, or they may discourage the change and make it more difficult.

Health promoting behaviors range from explicit, single-episode choices (e.g., pneumonia immunizations) to major life changes. Whether a given change is major or minor is individually determined. A major change is required when the behavior selected for alteration is deeply embedded in the individual's family and social life and represents challenges to his or her values and culture. A health promoting lifestyle is characterized by enduring participation in a set of behaviors. Health promotion, then, involves the reorganization of one's life pattern to exclude behaviors that interfere with health and include behaviors that enhance health. The core behavioral changes required involve the complex interplay of knowledge, motivation, and social support. The goal of health promotion interventions is to educate, motivate, and support people in their efforts toward healthful living.

Knowledge components of health promotion include behavior recognition, identification of alternative behaviors and the outcomes likely to result from different behaviors, and development of alternative patterns. Motivational aspects encompass values clarification and feedback about the extent to which one's current behaviors are congruent with one's espoused goals and values. Development of health promoting patterns requires that one actually engage in healthy behaviors. Social support and peer interaction play key roles in each of these. In fact, social support may convert a behavioral change, adopted for a few months or weeks under the supervision of a professional or the encouragement of a peer, from a brief alteration in a pattern to an enduring life process.

Social dimensions of behaviors that affect health are many and varied. Behaviors and habits developed over years of practice take on many meanings to individuals and to their social environment. The symbolic meanings of behaviors that influence health are complex. Consider, for example, how certain food aromas evoke pleasant, positive memories of family dinners or how the first nonsmoker in a group loses a key currency of interaction within that group.

Many individuals possess sufficient knowledge and motivation to adopt new behaviors or drop old destructive ones, but they are unable to meet the social costs of giving up an old familiar pattern. Social networks often affirm and reinforce the old, shared behavior pattern and may even punish new behaviors. However, people in the social network also can reward the new behaviors and help them become embedded in the individual's world. As behavior changes are initiated, connection with a group of people who are making or have made similar changes is often helpful. This group can affirm the challenges and desirability of the behavior change, provide advice, and emotionally support the person through the transition. Support from a group of people undertaking similar changes not only helps the individual to change health behavior, but also helps the individual to incorporate this change into his/her lifestyle by building supports within existing social networks or developing new supports.

The success of behavioral interventions in helping people to adopt health promoting behaviors has been demonstrated and the strengthening effects of social support documented. A review of studies designed to evaluate health promotion interventions concluded that interventions that combined emotional support with informational support had the strongest effects, and that support from lay counselors or peers was more effective than supportive interventions delivered by professionals (Cwikel & Israel, 1987). A more recent integrative review of descriptive studies found that self-efficacy and social support are the strongest predictors of a health promoting lifestyle (Gillis, 1993). Social support was effective in the initiation and maintenance of general health promoting practices (Zimmerman & Conners, 1989), exercise (Sallis, Hovell, Hofstetter, and Barrington, 1992), smoking cessation (Coppotelli & Orleans, 1985), and breast self-exam (Benoliel, 1994). Social support affects the acquisition of knowledge as well as behavioral changes (Benoliel, 1994; Hildingh, Segesten, Begtsson, & Fridlund, 1994). Examination of the effects of social support has, for the most part, been limited to interventions that enhance or increase the individual's face-to-face contacts. The potential for using computer technologies to insure the development of healthy lifestyles has not yet been fully exploited.

Computer technologies can play an important role in conveying knowledge for health promotion. Gustafson, Chewning, Bosworth, and Hawkins (1987), reviewing the literature on the impact of computer technology in health promotion, concluded that computer systems are effective vehicles for disseminating health promoting information. A key contribution of computer technology is the ability of computer systems to engage participants in the problem solving necessary to identify currently risky behaviors and appropriate new behaviors.

Computer networks, however, hold great promise not only for increasing access to the knowledge resources and motivational aides necessary for health promotion but also for providing a social environment likely to ensure the integration of behavioral change into an enduring life pattern. Like free-standing computer systems, computer networks serve as pathways to the life experiences of others as well as information. Computer networks also have the unique advantage of opening pathways for communication to either enhance existing social support or develop new support channels. Networks can provide access to one's existing support system or help one identify and connect with a new support network. This chapter examines the role of computer networks in providing the key social support necessary to increase individuals' engagement in health promoting lifestyles.

COMPUTERS AND SOCIAL SUPPORT

Many of the benefits of social support can be obtained through effective use of computer networks. Computer systems permit interested persons to review information, engage in interactive exploration and learning, and proceed at a pace deemed comfortable by the person. However, the limits of any educational intervention in terms of health promotion are also the limits of computerized health promotion educational processes.

In the Body Awareness Resource Network (BARN) project, high school students could access a library-based personal computer that held several programs addressing health promotion and youth. Topical content included sexuality, drinking behavior, self-concept, and peer relations. Evidence from the BARN project shows important risk-reduction behavior (Bosworth & Gustafson, 1991). Although this program was designed to promote solitary exploration of sensitive topics, investigators found that on more than 40% of occasions, two or more young people used the program together, interacting with each other while they interacted with the system. Thus, users created their own social environments that appeared to enhance the health promoting benefits of the system.

Computer networks are designed to link remote computers to each other, allowing sharing of common programs as well as access to communication pathways whereby users can interact with each other in a one-to-one or group fashion. Computer networks play an important role in health promotion in that they support peer interaction, provide access to informational and motivational programs that may be useful for health promotion goals, serve as a common meeting place for persons with shared interests, and enhance the ability of persons already linked in socially supportive relationships to maintain contact with each other.

Public bulletin boards, chat groups, and conferences that run on computer networks function in much the same way that face-to-face groups do (Ripich, Moore, & Brennan, 1992; Mihalo, 1985). Evidence suggests that the curative factors (Yalom, 1991) found in therapeutic groups also emerge in asynchronous interacting group situations. The factors of universality (finding others with similar experiences, problems or goals), re-creation of the family of origin (permitting re-examination of old behaviors and their meaning), and trust (the ability to try new behaviors in a safe environment) are particularly important.

In a year-long experiment called the ComputerLink, elder caregivers of persons with Alzheimer's disease (AD) had access to a specialized computer network that provided information, decision-making utilities, and communication services. Through computer terminals placed in their homes, caregivers accessed ComputerLink at a time convenient to them and chose from a variety of features. Features known to be successful in engendering social support, such as peer contact and information, were incorporated into ComputerLink. This section illustrates the value of projects like Computer-Link as social support for health promotion.

ComputerLink linked caregivers to each other and to a nurse moderator with the goal of assisting caregivers to provide home care. It consisted of three components: a communications module, an information module, and a decision-support module. The communications module included a Forum, where caregivers and the moderator publicly posted and read messages; a Question-and-Answer section in which subjects could anonymously post questions to a registered nurse moderator; and a private mail system. The nurse moderator facilitated social support through judicious observation and comment in the Forum discussion. The Electronic Encyclopedia (EE) provided over 200 indexed screens of information about AD, caregiving issues, community resources for AD, and self-care for the caregiver. The decision-support module guided caregivers through decisions using an analysis process that incorporated their own words and preferences, assisting them to make choices consistent with their values.

Offering social support via computers is a natural outgrowth of the increased use of technology in patient care. The data reported here are a subset of data gathered in a larger study evaluating the effect of Computer-Link on caregivers' decision confidence, decision skill, and isolation (Brennan, Moore & Smyth, 1995). In the larger study, 102 caregivers of AD patients were randomly assigned to either the ComputerLink intervention or a telephone comparison group. Each group contained 51 subjects. Four subjects withdrew from the computer group. Two individuals were not able to have the computer installed due to telephone-line difficulties, one had a health problem that precipitated withdrawal from the study before the computer was installed, and the fourth subject asked to have the computer removed from her home after the first week.

Data were collected on caregiver demographics and caregiver use of ComputerLink. Demographic variables were measured using an investigator-developed, structured interview guide. Forty-seven subjects had access to ComputerLink, 32 (68%) females and 15 (32%) males. The mean age of subjects was 60.3 years. The mean length of caregiving was 30 months. Relationship of the caregiver to the care recipient was primarily as spouse, daughter, or daughter-in-law. Most subjects had completed high school. Thirty percent of the subjects were working full time; 38% were retired. Ten (22%) of the households had children under the age of 18 and 7 (15%) had adult children living in the household. Thirteen (28%) of the caregivers also provided care for someone other than the person with AD. Only 3 of the subjects had experience using a computer in their home. However, nearly half stated that they had some prior experience with a computer, either at work, at school, or through a brief course in computer use. This chapter reports on subjects' uses of ComputerLink during a 1-week period occurring after all caregivers had been on the network for at least 1 month. Subjects read the Forum more frequently than they posted messages on it. During 1 week of the study, of the 80 accesses to the Forum, 14 messages were posted. As with a support group, caregivers used the Forum to share information, express their feelings and offer support to other members. Forum messages posted by subjects included information on caregiving issues, such as how to deal with wandering of the person with AD, or how to obtain diapers at wholesale prices. Of particular note, however, was the emotional content of the Forum messages. Caregivers typed messages expressing their feelings about mental and physical changes in the AD patient and their frustration regarding their caregiving situations. Many subjects expressed their gratitude at having a place to vent their feelings to "those who really understand."

ComputerLink use did not vary with education, race, length of time of caregiving, or relationship to care recipient. There was a significant difference in age, $t = -2.40$, $df = 45$, $p = .02$, between subjects who used ComputerLink during the study week and those who did not. The mean age of users was 56.5 years ($SD = 2.6$) versus 66.4 years ($SD = 32$) for nonusers.

SOCIAL SUPPORT VIA COMPUTERLINK

Antonucci and Jackson (1990) defined *social support* as interpersonal transactions involving affect (liking, loving, admiration, respect), affirmation (agreement, acknowledgment of appropriateness of action, statement or point of view), or aid (things, money, information, advice, entitlement). In the Forum, messages directed to the whole group or to individual users provided evidence of the interactive nature of social support. As indicated in Table 9.1, the content of the Forum messages during the analysis week provided evidence of these types of interpersonal communication.

ComputerLink had one explicit social support vehicle, the communications area. In addition, the encyclopedia and decision analysis offered options for social support (i.e., information support from the EE). Thus, ComputerLink provided social support in several ways: through substantive content present in the Electronic Encyclopedia, through motivational and values-clarification strategies in the decision-making area and through contact with peers and professionals. ComputerLink demonstrates that naive users can and will use a computer network to gain access to social support services. The social support benefits of ComputerLink may have been limited, however, by reliance on verbal or written communication, timing delays in sending and receiving responses, and altered social mores.

COMPUTERS, SOCIAL SUPPORT AND HEALTH PROMOTION

Computers have been shown to be effective devices for delivering health promotion, risk assessment, and clinical services in the community (Gustafson, Bosworth, Chewning, & Hawkins, 1987). Some of these computer applications are free-standing, special purpose programs such as Balance, a program created to plan balanced diets (Clark & Ellis, 1982). Others use

TABLE 9.1
Evidence of Social Support in ComputerLink Forum Messages

Aspect of Social Support:	Example from Forum Messages
Affect	
Liking:	I wish you all my good thoughts and prayers for the New Year.
Admiration:	Dear J: what a wise and thoughtful person you are. With all the problems you are facing, you always seem to have time to offer counsel and encouragement to others. I hope some day to have the privilege of meeting you in person. God bless you.
Affirmation	
Acknowledging appropriateness:	*Statement:* The hospice nurses have told me that my mom only has a few days left ... I feel so helpless. I have to get myself ready for this but I don't want to or I can't face what will be God's will. Thank you for letting me shed some tears.
	Response: It is hard to let go. My mother died when she was 85 years old, having suffered many strokes. N, I don't know how old your mother is but if she has suffered as long as my mother did, perhaps you should feel that now she will finally rest and be with God.... You will miss her ....
Aid	
Advice:	*Question:* I have been having trouble getting my husband to change his clothes at night so he can go to bed ... It is disturbing to have this reaction at bedtime because it upsets me so that I cannot go to sleep myself ... I feel so helpless sometimes and I think that there must be some way that I can handle this. What will I do when I have more serious problems? Well, I just had to vent my feelings tonight ... Thank you for being there for me to sound off.
	Answer: You may have to learn to lower your standards.... There are so many frustrating things about this disease that I feel if he wants to sleep in his shirt that is fine with me. I will give him a clean one the next day. It is no use getting so upset at bedtime that you cannot sleep.
	Response to Answer: Thank you for your suggestions. I think that I may have been trying to postpone as long as possible seeing my husband become content with a way of life different from what he has been accustomed to.... I see that others have had this problem and learned to live with it. I am sure that I will also.
Information	It is nearing the time for placement. If any of you have information about Brentwood Health Center, please let me know.... I would like to have a choice when the time comes. Please let me hear from you.

Note. From P. F. Brennan et al., *Western Journal of Nursing Research, 14*(5), pp. 669–670, copyright 1992 by Sage Publications, Inc. Reprinted by permission of Sage Publications, Inc.

telephone lines to make links from computers placed in homes. In an innovative early program, the Electronic Grandparent project linked elders in a nursing home with children in a day-care setting via computer terminals and telephone lines (Kerr & Hiltz, 1982). The Senior Net, funded by the Markle Foundation (Greenberger & Puffer, 1989), employed a computer network to provide elders with home-based interaction via an electronic network.

Educational programs delivered via computer software including CD-ROM and public access networks like America Online show promising results in conveying the knowledge necessary for health promoting lifestyles. The work of Gustafson and colleagues (1987) demonstrated that peer reaction and motivation for change can also be conveyed in this medium. Computer technology can aid in not only motivation, but can also aid an individual facing needs for control and scheduling. Diet planning programs, for example, provide a range of choices and then prescribe the scheduling of meals. The full impact of these systems, however, may be limited; education has been shown to be necessary but not sufficient for establishing enduring health promoting life patterns. It is possible that the difference is that education works either when there is adequate social support or when the behavior change is exclusively cognitive and not connected to the person's social support.

It may be impossible to disentangle the various components of social support provided within a computer network service. However, it may also be unnecessary. Research evidence (Cwikel & Israel, 1987) suggests that almost all successful interventions are delivered as some combination of emotional support, affirmation, and information. It is also possible that social support allows the explicit tailoring of static information that enables a person to integrate health promoting behaviors into his or her life. Social commentary on relevant information may be most powerful, and the computer network can deliver both information and the commentary.

HOW CAN COMPUTERS FACILITATE SOCIAL SUPPORT NECESSARY FOR HEALTH PROMOTING LIFESTYLES?

Computer networks, electronic links between computers at remote sites, are fast becoming an important communication strategy. Computer networks like CompuServe, America Online, the Internet, and proprietary networks open pathways to knowledge and educational resources and promote social interaction. Although computer networks provide exciting opportunities to access educational and motivational resources worldwide, we will focus here on their abilities to facilitate social interaction. Although the educational and motivational benefits are not inconsequential, they may be effective in this mode when they can more efficiently be linked with peer and professional social support.

Networks facilitate contact for information and peer support. They allow persons familiar with each other to maintain contact in the face of geographic or temporal barriers. In addition, because computer networks provide a place-independent space for meeting others with similar interests (Meyrowitz, 1985), they may serve to provide a new social context from which persons interested in health promotion can select peer supports.

Computer networks offer several advantages over face-to-face health promoting interventions. For example, computer network communication is written rather than spoken, and is asynchronous (that is, the sender and receiver need not be present simultaneously). This creates a situation of social anonymity, which in turn leads to less inhibited and more assertive communication. There is more equal participation, more uninhibited verbal behavior, and greater choice shift (Keisler, Siegel & McGuire, 1984; Zimmerman, 1987). Relationships formed over computer networks are as likely to reflect intimacy as those found in face-to-face encounters (Mihalo, 1985). Because the messages posted to the computer remain permanent, the content can be retained and accessed for long periods of time (Rafaeli, 1986). Computer networks may also be more convenient in that interested persons can access the system when they choose and do not need to travel to a central location to interact with others. Furthermore, because users may choose to read only, or to interact selectively with others, concerns about stigma and embarrassment may be reduced.

STRATEGIES FOR USING COMPUTER NETWORKS AS SOCIAL SUPPORT ENHANCERS FOR HEALTH PROMOTION

Computer networks offer many opportunities for tailoring social-support mechanisms. First, different routes can be taken to offer social support; these routes may be characterized as general or specific. General routes to social support involve encouraging persons interested in developing healthy lifestyles to use computer networks for interaction with others. Specific routes involve establishing and using computer networks organized around specific themes.

The Alcoholic Anonymous chapter that meets on CompuServe is an example of a specific route using a computer network to provide social support to enhance health promotion. Persons set up a section of a network organized around a specific theme and targeted for a special group. The strength of this approach is that it brings together people who share similar experiences and are striving toward the same goal. Research suggests that when a specific behavioral change is targeted, support that is specific to that change is more effective than perceived general support (Cwikel & Israel, 1987). Measures of general social support, however, have been consistently associated with measures of overall healthy lifestyle patterns (Palank, 1991). This may be because of the effects of social support on the persons' belief that they will be able to make behavioral changes (Kelly, Zyzanski, & Alemango, 1991). In addition, there is some evidence that changing one set of behaviors may lead to changes in other behaviors. For example, a person who begins a regular walking program often chooses later to alter smoking and eating behaviors and may enhance his or her ability to handle stress. Reliance on specific routes to a narrowly targeted behavior may be too limiting to ensure enduring life patterns. It remains to be seen whether the specific route is advantageous initially to engage a person in a specific health promoting behavior.

General routes, or multipurpose networks, have the advantage of offering interested individuals a package of support services. If we agree that the desired goal of health promotion interventions is lifestyle change, not isolated behaviors, it seems logical to rely on more general routes to providing social support. Given the initial draw that specific, focused support networks offer, we might find greater success if general networks are organized around engaging themes, such as developmental groups or people facing similar life challenges.

A second option in employing computer networks for social support directed toward health promotion is to establish a focused group for a specific set of individuals or address health promoting interventions to members of existing networks. In the first case, we create a new social support network, linking individuals who do not know each other but who have common goals and common needs. In the latter case, we target interventions to existing users. A modification of this is to use an existing network to connect a person naive to computers but in need of social support.

Health promoting behaviors range from explicit, single-episode choices to major life changes. Health promotion requires knowledge, motivation, social support, and behavioral change. Social support plays a key role in insuring that behavior changes become an enduring lifestyle pattern. Computer systems provide individuals with access to both new information and peer support; necessary components of initiating and maintaining behavioral changes. ComputerLink, an experimental computer network, was accessed by ill persons and home caregivers and found to be an effective and efficient vehicle for providing social support to home-bound persons. Computer technologies hold great promise as tools for promoting social support and health lifestyles. The rapid growth in computer systems in the homes, in public settings such as libraries and schools, and in the workplace facilitates easy access to health promotion literature and to the support of interested others who can assist an individual in attaining high-level wellness.

REFERENCES

Antonucci, T. C., & Jackson, J. S. (1990). The role of reciprocity in social support. In B. R. Sarason, I. G. Sarason, & G. R. Pierce (Eds.), *Social support: An interactionist view* (pp. 173–198). New York: Wiley.

Benoliel, J. Q. (1994). Effects of education and support on breast self-examination in older women. *Nursing Research, 43*, 158–163.

Bosworth, K., & Gustafson, D. (1991). CHESS: Providing decision support for reducing health risk behavior & improving access to health services. *Interfaces, 21*, 93–104.

Brennan, P. F., Moore, S. M., & Smyth, K. A. (1992). Alzheimer's disease caregivers' uses of a computer network. *Western Journal of Nursing Research, 14*, 662–673.

Brennan, P. F., Moore, S. M., & Smyth, K. A. (1995). The effects of a special computer network on caregivers of persons with Alzheimer's Disease. *Nursing Research, 44*(3), 166–172.

Clark, K., & Ellis, L. (1982, October). *Balance: An interactive program for the design of balance study diets.* Paper presented at the Symposium on Computer Applications in Medical Care, Los Angeles, CA.

Coppotelli, H. C., & Orleans, C. T. (1985). Partner support and other determinants of smoking cessation maintenance among women. *Journal of Consulting and Clinical Psychology, 53*, 455–460.

Cwikel, J., & Israel, B. A. (1987). Examining mechanisms of social support and social networks: A review of health-related intervention studies. *Public Health Reviews, 15*, 159–193.

Gillis, A. (1993). Determinants of a health-promoting lifestyle: An integrative review. *Journal of Advanced Nursing, 18*, 345–353.

Greenberger, M., & Puffer, J. C. (1989). Telemedicine: Toward better health care for the elderly. *Journal of Communication, 39*(3), 137–144.

Gustafson, D. H., Bosworth, K., Chewning, B., & Hawkins, R. P. (1987). Computer-based health promotion: Combining technological advances with problem-solving techniques to effect successful health behavior changes. *Annual Review of Public Health, 8*, 387–415.

Hildingh, C., Segesten, K., Begtsson, C., & Fridlund, B. (1994). Experiences of social support among participants in self-help groups related to coronary heart disease. *Journal of Clinical Nursing, 3*, 219–226.

Kelly, R. B., Zyzanski, S. J., & Alemango, S. A. (1991). Prediction of motivation and behavior change following health promotion: Role of health beliefs, social support, and self-efficacy. *Social Science Medicine, 32*, 311–320.

Keisler, S., Siegel, J., & McGuire, T. W. (1984). Social and psychological aspects of computer mediated communication. *American Psychologist, 39*(10), 1123–1134.

Kerr, E. B., & Hiltz, S. R. (1982). *Computer mediated communication systems.* New York: Academic.

Meyrowitz, J. (1985). *No sense of place: The impact of electronic media on social behavior.* New York: Oxford University Press.

Mihalo, W. E. (1985). The microcomputer and social relationships. *Computers and the social sciences, 1,* 199–205.

Palank, C. (1991). Determinants of health-promotive behavior: A review of current literature. *Nursing Clinics of North America, 26,* 815–832.

Rafaeli, S. (1986). The electronic bulletin board: A computer driven mass medium. *Computers and the Social Sciences, 2,* 123–36.

Ripich, S., Moore, S. M., & Brennan, P. F. (1992). A new nursing medium: Computer networks for group intervention. *Journal of Psychosocial Nursing, 30*(7), 15–20.

Sallis, J. G., Hovell, M. F., Hofstetter, C. R., & Barrington, W. E. (1992). Explanation of vigorous physical activity during two years using social learning variables. *Social Science and Medicine, 34,* 25–32.

Yalom, I. D. (1985). *The theory and practice of group psychotherapy.* New York: Basic Books.

Zimmerman, D. P. (1987). A psychological comparison of computer-mediated and face-to-face language use among severely disturbed adolescents. *Adolescence, 22*(88), 827–40.

Zimmerman, R., & Connor, C. (1989). Health promotion in context: The effects of significant others on health behavior change. *Health Education Quarterly, 16,* 57–75.

10

Creating Illness-Related Communities in Cyberspace

Dirk Scheerhorn
Philadelphia College of Pharmacy and Science

In 1890, Cathell in *The Physician Himself* wrote of the perils of an informed and educated clientele; he warned that educating patients would cheat doctors out of further visits or other potential patients. In his words:

> Especially avoid giving self-sufficient people therapeutic points that they can thereafter resort to …. It is not your duty to cheat either yourself or other physicians out of legitimate practice by supplying this person and that one with a word-of-mouth pharmacopoeia for general use.

Since the time of Cathell, the model of the wise physician with ownership over medical knowledge has flourished. This ownership contributed to the status and authority of physicians over patients (Starr, 1982). Although cracks occasionally appear in the model, there remain societal, institutional, cultural, and even communicative impediments to change. Medicine, at least in our lifetime, has always been more interested in providing the fish than teaching the art of fishing. Or, as Kreps (1996) suggested "the modern health-care system is designed to meet the needs of health-care providers and not consumers" (p. 42).

However, new technologies, along with other factors such as the corporatization of healthcare, threaten to weaken the ownership of information among an elite and may contribute to the undoing of medicine as we know it (see also chapters 11 to 13 this volume). How might a new technology—in this case, a private text-based messaging system among an illness-related group—contribute to the changing face of medicine? This chapter probes that and related questions. It does so through experiences and conclusions gleaned from a project known as the Hemophiliacs In Good Health Network (HIGHnet). In essence, the chapter represents both the highlights and lessons learned from the HIGHnet story.

The following path is traversed in the telling of this story: Because no system—in this case, an electronic network of an illness-related group—can be separated from its larger sociopoliticocultural milieu, the illness of hemophilia and its wider community dynamics are first overviewed. The needs of the hemophilia community and presumably other illness-related communities are then described. Next, prior research on interactive systems of use by illness-related groups is briefly reviewed. The HIGHnet experience, although described in detail elsewhere (Scheerhorn, Warisse, & McNeilis, 1995), is then summarized. Various aspects of the HIGHnet system, including its sites, users, and design are outlined as well as its utilization, functions, and impact. Finally, future research questions are posed concerning HIGHnet and similar systems designed for illness-related communities.

HEMOPHILIA: AN OVERVIEW OF THE ILLNESS AND THE COMMUNITY

Hemophilia represents a class of genetic diseases that prevents the production of proteins necessary for normal coagulation. The most frequent type of hemophilia, hemophilia A or factor (protein) 8 deficiency, is found in 1.25 of every 10,000 male births. Hemophilia A is almost always found in males although a related condition, Von Willebrand's disease, has equal incidence among men and women. As a chronic illness, hemophilia is characterized by both routine and traumatic bleeding into the joints, muscles, and organs. A variety of factors contribute to the frequency and severity of bleeding episodes and, although genetic therapies loom, there is no known cure for hemophilia.

Until recently, people with hemophilia could expect severe crippling and a shortened life expectancy. However, during the 1960s and 1970s, factor replacement products and home infusion therapy greatly improved life and instilled much normalcy for the person with hemophilia. Unfortunately, factor replacement products are made from human blood donations, and, unlike a transfusion in which one or some small number of donors are implicated, a dose of factor will generally contain the blood of 10,000 donors. Predictably, although sadly too late, in the early to mid-1980s, factor was found to contain the HIV virus. The most tragic of all ironies; the identical treatment that gave life to so many was, and is now, taking it away. At one time, an estimated half of the hemophilia population was HIV-positive ("*American tragedy*," 1994), although that percentage drops with deaths and new births. Although more purified factor products and HIV-screening methods developed in the mid-1980s prevent the problem for most young hemophiliacs, the damage to the hemophilia community is unparalleled and multilayered, affecting individuals, families, groups, and organizations.

In addition to the loss of life suffered in the community, blame, guilt, and conflict are recurrent among the primary groups and institutions in the community depicted in Fig. 10.1. (Individuals in the community are indicated by circles, whereas institutions are squares.) Although the picture attempts to show relationships and connections, it fails to adequately demonstrate the interwoven dynamics wherein each part influences, and is influenced by, each and every other part of the community.

At various times and in various ways, these parts of the hemophilia community have been adversaries—often in courts of law. Through the turmoil and tension, however, the community endures. Hemophiliacs continue to treat themselves using replacement therapies, genetic therapies are in development, home-care companies and hemophilia treatment centers compete to survive amid managed care, and most facets of the community work toward federal compensation for lives lost.

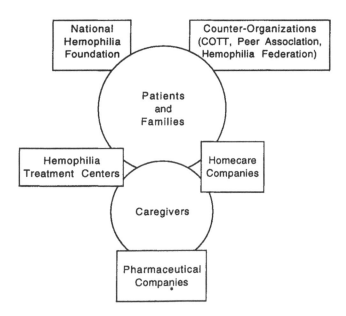

FIG. 10.1.   Parts of the hemophilia community.

THE NEEDS OF THE HEMOPHILIA COMMUNITY

People with chronic illnesses, at least those within the hemophilia community, have several needs beyond the average person. Improvements in the quality of life for persons with hemophilia will presumably follow as the needs described next are satisfied. Importantly, the needs interact with one another such that improvements in one should foreshadow improvement in the others.

Good Medicine, Good Physician. Probably the most important need of the chronically ill person is sound medicine practiced by a skilled specialist. Finding a skilled specialist for the person with hemophilia can be difficult because at a minimum, a hematologist is needed and, even better, a physician specializing in hemophilia. However, because of the low incidence rate of hemophilia, these sorts of physicians are rare and choice for the patient may therefore be limited. Obviously, the problem is especially acute in rural areas and small towns.

Finding the appropriate expertise, or finding choice in expertise, is generally accomplished in at least two ways: physician and patient referral. Formally, many patients find their hematologist or hemophilia specialist (who often practice at hemophilia treatment centers) through their family doctor. Similarly, referrals are made by the hematologist to the orthopedic surgeon and/or to another hematologist to treat joint and liver problems associated with hemophilia.

Informally, referrals are made by other patients in the hemophilia community. These referrals often involve consideration of bedside manner although medical competence is also assessed and shared. Of course, the more knowledgeable the patient concerning the medical aspects of the illness or its complication, the more equipped that patient is in assessing physicians and making wise referrals.

For many people, the choice of physician is relatively unrestricted and cavalier; for someone with a rare chronic illness, the choice is highly constrained and vital to good health. Making known all choices for competent physicians can only improve the health of the community.

Good Information, Good Education. As mentioned earlier, the person who is uninformed or unknowing about his or her chronic illness has fewer means to make distinctions among physicians. Just as important, however, is that the same person also has fewer means to make judgments about the medical treatment she or he is receiving.

Whenever we proclaim the virtues of choice, we must first prepare the audience for choice. In this case, that audience needs to understand the illness, know its causes and complications, and know various treatment options and probable outcomes. Knowing this information, the patient is prepared to work with a particular physician on the appropriate management of the condition. Improved health stems from the involvement of an informed and knowledgeable patient.

Many sources of information exist concerning most chronic illnesses. Physician and/or other caregivers, illness-related groups or organizations, peers, and the public library all would prove useful in the patient's acquisition of knowledge concerning a particular condition.

Peer Support. Beyond the generalized need for social support that may improve quality of life, the person with a chronic illness has a particularized need to be linked to peers within the illness-related community. Each rare or chronic illness is unique, characterized by personal and private nuances and subtleties. Because the complexity of the illness makes it unlikely that outsiders will have new or useful information, the closer one individual's circumstance needs to be to the other in a relationship in order to derive more benefits from the link. A parent of a child with severe hemophilia will benefit greatly from interaction with another parent of a child with severe hemophilia. Closeness in circumstances affords greater communicative resources for the individual seeking and providing support. Similar circumstances make stories more relevant, information more credible, and empathy more likely.

These peer linkages are important not only as sources of social or emotional support, they also may fulfill other needs. For example, peers typically are more open with one another when discussing physicians and their medical prowess than they are with other health care professionals. Peers are able to relate through stories of similar experiences. Peers are able to advise on shortcuts through and around institutional and bureaucratic barriers (e.g., insurance policies, hospital admissions).

Locating peers within illness-related communities can be more difficult than one might imagine. Privacy may prevent both professionals and organizations from releasing names. In the case of hemophilia, the low levels of incidence may make finding a peer in reasonable proximity impossible. Meetings and educational symposia are sometimes offered although these tend to center around large population centers, making it again difficult for someone in a remote area to access. Moreover, most public fora are more concerned with disseminating generalized information than with matching particularized patients with one another.

Economic Support.   Although all desire affordable health care, the person with chronic illness has an acute need for economic support; unfortunately, the practices of health insurance companies make those with chronic illness the least likely to have it. Thus, at least two avenues of attack are available for the chronically ill. The first involves individual action. Whether through cost-saving techniques, shopping around for low-cost products and services, or seeking governmental or private assistance, the individual can initiate the quest for lowered health care costs. A second course of attack concerns collective action. Insurance laws, access to government-supported clinical trials, and research agendas are created and controlled by legislative bodies. Working as collective groups and organizations, policies and practices can be changed in state houses and in Washington. Notice that both of these action strategies are actualized through peer connections and the dissemination of information concerning cost-related matters.

The items of needs discussed usually do not occur in isolation from one another. Efforts to satisfy one need will in many ways serve to accomplish other needs. For example, good information will make the selection of a good physician more likely and may also lead the patient to avenues of economic support.

CYBER NETWORKS
AMONG ILLNESS-RELATED GROUPS

Recent advances in interactive technologies provide new channels and forums for communication about a vast array of topics and problems, including disease and illness. As a result, computer-based networks among illness-related groups are becoming increasingly popular as a means for information and social support (Lamberg, 1996). Research exploring the development, use, and functions of illness-related groups, on the other hand, remains limited.

To date, the work of Brennan and her colleagues using the Cleveland Freenet has provided the most systematic exploration of Cyberspace networks for people with a similar health condition (Brennan, 1992; Brennan, Moore, & Smyth, 1991; Brennan, Ripich, & Moore, 1991: Ripich, Moore, & Brennan, 1992). The research interests expressed in this work also most closely approximate those of the HIGHnet project.

One set of studies has focused on AIDS patients and their caregivers (Brennan, 1992; Brennan et al., 1991; Ripich, Moore, & Brennan, 1992). These studies investigated whether computer network usage on the Cleveland Freenet by persons with AIDS would show a reduction in feelings of social isolation, improve decision-making skills, provide health information, and offer peer support. After controlling for depression, results suggested that usage by participants was significantly related to reduced feelings of social isolation. In addition, respondents reported communication as their primary reason for accessing the network. Brennan (1992) concluded that these types of networks were used primarily by patients to exchange stories and ask for advice from others and that this use fulfills an especially great need for those with mobility problems or those who are hospital bound.

Two other studies (Brennan, Moore, & Smyth, 1991, 1992) examined the use of a network by caregivers of persons with Alzheimer's disease. Again, the system proved effective for information-sharing and story-telling. Based on the explicit statements of support and encouragement among the caregivers in their messages, the researchers concluded that the network was a success. Although the extant literature is limited, it is at least suggestive that cyber connections in illness-related communities are possible and potentially valuable.

HIGHNET: A NEW TECHNOLOGY
LINKING THE HEMOPHILIA COMMUNITY

In April 1994, HIGHnet was launched and remained active until early 1995. (A revised HIGHnet2 came online in January 1996.) Unlike most other online health networks, HIGHnet was created exclusively for members of an illness-related group—browsers or surfers could not access it. Only members of the wider community were given access to its use. Its users provided information and support to one another—ownership and control of content was not individual-or institution-based. The network was developed for two traditional media functions; namely, to provide an avenue of *communication* among members of the community and to provide *access to information*. Directly or indirectly, both functions are suspected of improving the well-being of ill people.

The hemophilia community was well-suited for this sort of network for a number of reasons. People with hemophilia have physical limitations, are widely scattered throughout the population, experience psychosocial trauma (particularly with the threat of HIV), and are a relatively stable and accessible population (Scheerhorn et al., 1995).

Sites, Users, and Design Features of HIGHnet

A total of 58 individuals at 34 home sites and two hemophilia treatment center (HTC) sites were enrolled into the network over approximately a 4-month period. The participants varied greatly in their ages (7 to 71 with an average age of 36). Similar diversity was reflected in sex (33 men and 25 women), education (less than high school to post graduate), and annual income (less than $10,000 to more than $60,000). The locations of the sites were widespread across the state of Ohio and northern Kentucky with a plurality in Columbus, OH. In order to accommodate those users that lived outside of the Columbus vicinity (where the network is based), an 800 phone number was obtained. Thus, for local as well as long distance callers, no cost was incurred by users.

In an effort to achieve user diversity, computer equipment was provided to some families. Twelve of the sites were provided a computer and modem, 7 a modem, and the remaining 15 sites needed no additional equipment. Three kinds of training took place across the 58 individuals. Thirty-seven received face-to-face/hands-on training, 9 watched over the shoulder of another family member, and 12 learned the bulletin-board system on his or her own time or from another family member. All sites were provided communications software at no expense and offered phone support to answer questions related to the system.

Types of users on HIGHnet fell into six categories relating to hemophilia; 22 were adult hemophiliacs, 16 were parents, 9 were spouses, 3 were HTC nurses, 2 were HTC social workers, 4 were other family members, and 2 were young hemophiliacs. Based on their typology, users were assigned to various conferences found on the HIGHnet bulletin board. Conferences described areas of the bulletin board dedicated to particular subgroups or various topics. In a very real sense, subgroup conferences were designed as social support or self-help groups where membership was limited to those sharing a common problem (Cline, 1990). Access into these conferences depended on the individual and their connection to the person with hemophilia. For example, only adults with hemophilia have access to the messages that are posted in the Adults With Hemophilia conference. In many ways, conferences were viewed as rooms and access to the rooms was limited depending on the keys (menu options) provided to each of the participants. Other subgroup conferences existed for spouses of adults with hemophilia, parents of a child or adult with hemophilia, young adults with hemophilia, and siblings of a person with hemophilia. HTC staff were not given access to any of the group-defined (adult with, spouse, etc.) conferences. All network members had access to the HTC conference and the All

Network Members conference. In addition, two conferences were later established around special topics and were available to all network members. One concerned legislative news, the other a forum for a statewide hemophilia service organization. Interestingly, these topical conferences exhibited a second (in addition to sharing) characteristic of social support groups, namely project work (Radley, 1994). Talk in these conferences often concerned the planning of meetings, task assignments, and other means toward organizing collective action.

Each user also could send private mail to any other single member. Aside from these differences in options within the conference menu, the system was identical for all users.

Utilization and the Functions of HIGHnet

By the 18th week of the project, a total of 2259 messages had been sent among the various conferences, and 58 network members were connected to the system. Users sent from a low of 1 message to the HIGHnet to a high of 346 postings. The mean number of calls to the bulletin board per user per day was .42. Mean number of messages sent per call was .38. And the mean number of minutes spent per call by each user was 7.5.

Results indicated that use (number of postings sent) was significantly related to the type of training the participant was given. Persons with more individual and face-to-face training utilized HIGHnet more. On the other hand, no other factor including demographics (age, income, education, sex), type of user (person with illness, spouse of, caregiver of, etc.), or the site's equipment needs influenced use (Scheerhorn et al., 1995).

Postings, both private and public, left on the HIGHnet were studied for the specific functions they seemed to serve. A brief description of the six functions follows (see Scheerhorn et al 1995, for more details).

Education. Although systematic coding has not yet taken place, thus making frequencies impossible to report, most postings had some educational component to them. Usually, the education was packaged in a story ("I remember when that happened to ... ") and concerned some aspect of treatment, often its consequences. Also, as in the following example many times the posting was education about education; where or how to seek more information about a particular topic or problem.

Hi <first name>!

My name is <first name> and I have a <description> named <first name> who is severe F VIII. I just went and looked up information on Factor IX products in the April '94 issue of the *Parent Exchange Newsletter*. If you are not getting that newsletter I would suggest that you do. It is free and written by the mom of a young child with hemo. The only Monoclanal factor IX on the list is MONONINE made by ARMOUR. I am not sure if there is any research being done on a recombinant factor IX. It appears that your child is receiving the best medication available.

IMHO (in my humble opinion)

Outreach.   This entails widening the sphere of, or participation by, community members. Perhaps HIGHnet's greatest success was the case of a mother and son (with hemophilia and HIV) living in rural Ohio, barely able to type, who both became very heavy users. Neither had spent any significant time with other persons or families with hemophilia until HIGH-net. As a consequence of their interactions on HIGHnet, they became friendly with other community members. Both later attended two educational symposia dedicated to hemophilia and HIV.

Advocacy.   Organizing collective action was another specific function of the network. As suggested earlier, illness-related communities have a special interest in laws, rules, and regulations germane to their illness. Many times, collective action can positively influence those policy issues. HIGH-net served as a sort of campaign central for state and national legislative initiatives. The following postings, sent to all, concerns the sharing of names and addresses and a prompt to act.

> TO WHOM IT MAY CONCERN, I AM TRYING TO CONTACT <alias> TO GET MORE INFORMATION ABOUT WRITING TO CONGRESS-MAN AND SENATORS. I HAD THIS INFORMATION, BUT HAVE MISPLACED IT AND WOULD WRITE IF I HAD INFO. PLEASE HELP NOW. I THINK THIS IS VERY IMPORTANT AND WOULD URGE EACH AND EVERYONE OF YOU TO WRITE LETTERS.
>
> BYE FOR NOW, <signed alias>

Improved Psychological and Physical Health.   Anecdotally, several postings, including this one, directly suggested that HIGHnet made an individual feel better.

> I just want to let you know once again how much Highnet has changed my life for the good. <first name> will be able to attend tomorrow's meeting with her brother because I will babysit <first name>. Without Highnet, this wouldn't been possible because I wouldn't have been aware of the need. This

has strengthened our friendship, hell it created that friendship. Can't thank you enough for hanging in there with this project and you've really hit big time with it. Let me know if you need any commentary etc. Where have you been? My hives are gone and so are the anti-depressants. Oh well, I tried the drug of the 90's. I'll just have to get drunk more often. Dumped <first name> for the time being and I feel better already. I guess I didn't realize how really "down" she made me feel. Gotta go clean. Can't let <first name> and <first name> see my house like this—its sooo bad! Take care. Miss hearing from you <signed nickname>

Business. A very common function within the HTC conference is the performance of business, including appointment setting and confirmation. An example follows:

HI <first name>, WE HAVE LOST TRACK OF YOUR APPOINTMENTS AND WERE WONDERING WHEN YOU WOULD BE COMING IN TO SEE DR. <name>.? JUST LET US KNOW. HOW DO YOU LIKE THIS BULLETIN BOARD? I THINK IT IS PRETTY NEAT—YOU'LL HAVE TO TELL <dr. last name> ALL ABOUT IT, I'M SURE SHE WOULD FIND IT VERY INTERESTING.

<signed first name>

The business function was also apparent in the community's planning of an educational event.

Cost Savings. The final function of HIGHnet is perhaps its most powerful. One of the unique features of illness-related cyber communities may be the "publicizing" of otherwise private experiences. In the case of HIGHnet, the cost of a treatment that a consumer had negotiated was revealed by that consumer in a posting on HIGHnet. Two other consumers, having learned of the negotiated cost, were able to get the same cost for themselves. A conservative estimate for a 1-year period demonstrated cost savings to the patients (and/or the insurance companies) of $25,000. HIGHnet, in the grand ledger, had paid for itself.

Impact

In a very real sense, HIGHnet was able to address each of the four needs of the chronically ill discussed earlier. The overall impact of HIGHnet is best summarized by three levels of analysis: individual, organizational, and community. Individually, HIGHnet was felt in a number of ways. Some individuals benefitted financially, some were able to improve their health, some were able to conduct some business, some were able to learn more about their illness and its treatment, and most were able to increase and strengthen friendships. Organizationally, HIGHnet was able to facilitate planning of HTC appointments and news, and a volunteer service organization's educational symposium for patients/consumers. The community benefitted by greater and improved advocacy, outreach, and an almost indescribable pride in the communication created by the system. In many conversations within the community, both online and off, HIGHnet *is* the message. Its members share something beyond hemophilia, and, in turn, the bond of hemophilia becomes qualitatively different.

FUTURE RESEARCH

Whether or not systems like HIGHnet will vastly and positively transform the practice of medicine and the experience of the ill is not yet clear. Many questions remain unanswered and in some cases unasked.

These questions concern both matters of theory and policy but are, in part, undergirded by serious and difficult questions of value. The experience of HIGHnet focuses the research agenda on the following issues.

Demonstrating Outcomes

The most fundamental research needed is the systematic analysis of messages/postings and their short and long term effects on health-related beliefs and behaviors. Two strands, one concerning the unitization, description, and meaningfulness of messages, and a second on whether and how communicative actions enhance an individual's health and well-being, must merge in the demonstrations of *outcomes*. Although the need for more outcome-based research is present across the health communication spectrum (Kreps, O'Hair, & Clowers, 1994), we know especially very little about the health consequences of various forms, contents, and usages of cyber postings.

Telemedicine or Telehealth: Who Benefits?

The reason outcomes-based research is so vital concerns money and the allocation of scarce resources. Simply put: Who is going to pay for the establishment and maintenance of these systems?

One of the risks this project faced, and attempted in some ways to overcome, was of further widening the gap between the information haves and the information have-nots. Kroger (1994) suggested this is accomplished by establishing systems to include rather than exclude the underserved. Through the use of low-end technology, a toll-free access number, and face-to-face training, HIGHnet brought social support, including health information, into the homes of the poor and the undereducated.

Closely related is the larger problem of multiple visions or, who benefits most from which new interactive technologies? A recent experience crystallizes the issue. When describing HIGHnet to a group of very prominent and relatively consumer-friendly hemophilia physicians, one became extremely concerned about the potential for information abuse. Who makes sure the information and advice is correct?" he asked. The explanation that other patients, nurses, and even doctors were all welcome to contribute did little to dissuade his concerns. However, when the idea emerged that hemophilia physicians might have a private conference in which to share information, he became most supportive, even excited. The problem is a classic one: Those with resources (money, information, education) are more apt to access and utilize new systems of communication and information, and yet, they are least in need. Those most in need, unfortunately, are not likely to have a computer nor the knowledge and experience necessary to use one.

Fortunately, the crack in the control over medicine created by managed care provides an opportunity for the establishment *and* financing of new communication systems *if* positive outcomes can be demonstrated. If improvements in health as a result of participation in cyber communities can be documented, insurance providers are likely to pay for them. Again, the agenda points in the direction of outcomes but also suggests an additional step in the research program. In an environment of scarce resources, communication researchers need to compare their cost/benefit analyses against other medicinal therapies.

Accuracy and Ownership: Bumps on the Highway

A final question refers back to the comment made earlier by the skeptical physician. He raised an extremely important dilemma for community owned systems such as HIGHnet. Who is responsible for inaccurate information and who is accountable for its potential negative consequences?

In the case of HIGHnet, accuracy was left only to its members and (we hoped that) no one was legally accountable. Are systems like HIGHnet, composed primarily of consumers/patients, capable of maintaining standards of accuracy without accountability? Are patients/consumers capable of collecting and disseminating information equal in quality to, and in the eyes of, physicians? How can patients and physicians work together to insure high-quality information and how can systems be designed and developed to facilitate those workings.

Questions and problems concerning the accuracy of information and accountability in cybersystems will, in part, be played out in courts of law and legislatures. However, communication researchers have a role in those policy discussions. Policy should be informed by experience, and communication researchers are best equipped to interpret and comment on the communicative experiences of cyber communities.

CONCLUSION

In this chapter, the story of the HIGHnet project was summarized. The story is only a part of the ever-changing technological and health environments in which we live. New technologies afford unparalleled opportunities for individuals to interact and learn from one another. New healthcare practices place ever-increasing emphasis on lowering costs while decreasing the power of physicians. Whether and how illness-related cyber communities develop in these new environments is highly speculative although developments should be informed by health communication theorists and researchers.

ACKNOWLEDGMENTS

This project was made possible through the generous support of Quantum Health Resources of Orange, California, and the Centers for Health Policy Studies and Advanced Study in Telecommunication at Ohio State University.

REFERENCES

An American tragedy in Iowa. (1994, February 7). *Newsweek, 123*, 44–45.

Brennan, P. F. (1992). *Computer network homecare support with AIDS: A randomized sample.* Unpublished manuscript, Case Western Reserve University, Cleveland, OH.

Brennan, P. F., Moore, S. M., & Smyth, K. A. (1991). ComputerLink: Electronic support for the home caregiver. *Advances in Nursing Science, 13*, 14–27.

Brennan, P. F., Moore, S. M., & Smyth, K. A. (1992). Alzheimer's disease caregivers' uses of a computer network. *Western Journal of Nursing Research, 14*, 662–673.

Brennan, P., Ripich, S., & Moore, S. (1991). The use of home–based computers to support persons living with AIDS/ARC. *Journal of Community Health Nursing, 8*(1), 3–14.

Cathell, D.W. (1890). *The physician himself.* Philadelphia: Davis.

Cline, R. J. (1990). Small group communication in health care. In L. B. Ray, & L. Donohew (Eds.), *Communication and health: Systems and applications* (pp. 69–91). Hillsdale, NJ: Lawrence Erlbaum Associates.

Kreps, G. L. (1996). Promoting a consumer orientation to health care and health promotion. *Journal of Health Psychology, 1*(1), 41–48.

Kreps, G. L., O'Hair, D., & Clowers, M. (1994). The influences of human communication on health outcomes. *American Behavioral Scientist, 38*(2), 248–256.

Kroger, F. (1994). Toward a healthy public: Models, messages, and meaning. *American Behavioral Scientist, 38*(2), 215–233.

Lamberg, L. (1996, April 1). Patients go on-line for support. *American Medical News, 39*,(13), pp. 10–12.

Radey, A. (1994). *Making sense of illness: The social psychology of health and disease.* Thousand Oaks, CA: Sage.

Ripich, S., Moore, S., & Brennan, P. (1992). A new nursing medium: Computer networks for group intervention. *Journal of Psychosocial Nursing, 30*(7), 15–20.

Scheerhorn, D., Warisse, J., & McNeilis, K. S. (1995). Computer-based telecommunication among an illness-related community: Design, delivery, early use, and the functions of HIGHnet. *Health Communication, 7,* 301–325.

Starr, P. (1982). *The social transformation of American medicine.* New York: Basic Books.

III

THE FUTURE
OF INTERACTIVE TECHNOLOGY
FOR THE PROMOTION
OF HEALTH

11

Digital Interactive Media and the Health Care Balance of Power

Gary Kahn
Healthbridge Systems

As I explore the role of interactive media in health promotion, it is important to consider the broader context of the systemic changes coming to health care as a result of this and related technologies. The focus of this chapter is to look at how interactive media will affect the healthcare balance of power in the years to come. *Power*, for my purposes, refers to control of decision making related to access to, and utilization of, the resources necessary to effect desired health outcomes. *Balance* refers to the relative weight given to the priorities of the various players in the equation—consumers, providers, payers, regulators—both in their institutional aggregations (e.g., medical practices, hospitals, managed care organizations, unions, insurance companies, licensing boards) and the individuals acting on behalf of these entities. The forces shaping the future of health care are shifting power away from the individual provider and institution toward administrative components of emerging managed care megasystems; the future, however, holds the promise of a further dramatic shift toward the consumer.

This chapter examines the shift away from a provider-centric to a consumer-centric delivery environment and the concomitant shift of emphasis away from professionally directed illness care toward that of self-care, prevention, and health promotion. In addition to social, economic, and cultural factors, a core ingredient in this shift in the health care balance of power is information technology. As we transition from the Industrial Age to the Information Age, an understanding of how information and communication technologies are enabling the transformation can help us predict and perhaps influence possible paths we might be taking.

These transitions will be examined in four sections: In *The Paradigm Shift,* some of the transformations that are occurring or likely to occur in the health care landscape will be identified. Next, an analysis of the *Power Shift* taking a brief glance at historical, current, and projected roles in determining health care decision making. Following this, a few key *Enabling Technologies* are reviewed to show how their coalescence into Health-Oriented Electronic Performance Support Systems will set the stage for some dramatic changes that are to come. Finally, I speculate on *The Path* or trajectory the changes might take, providing one scenario to dramatize the role of the new interactive technologies in my vision of the emerging consumer-driven health care marketplace.

THE PARADIGM SHIFT

In the latter years of the 20th century, health care in the United States has been characterized more by its lack of philosophical and economic underpinnings than by a coherent set of goals and administrative systems. We find ourselves in this state of turmoil due to a number of dramatic transforming forces and trends. This transformation is often referred to as a *paradigm shift* because the changes in the system are so fundamental that we must throw out many of our previously held assumptions and beliefs. In order to appreciate the role that information and communication technologies are playing, it is important to consider some of the more important social, political, and economic transformations that are taking place in health care. Three major arenas that bear examination include significant changes in the goals of our health care system, the redefinition of community, and our human limitations in an increasingly complex health care environment.

Goals of Health Care—Toward a Broader Definition

Our current system of health care delivery is optimized to provide personal health services primarily when illness has become apparent. For most of the last half-century in this country, our focus has been on narrow physical parameters of health, often minimizing or ignoring key ingredients of the broader definition of *health* accepted by the World Health Organization more than two decades ago—a definition that includes emotional, social, spiritual, and economic well-being. As we turn the corner into the new millennium, we are beginning to see some chinks in the armor of the status quo—even as if recent health care reform debates remind us we have a way to go.

There is now a perceptible shift in our health care system away from an illness-care orientation toward an increasing focus on prevention, health promotion, and the importance of community in achieving individual and societal health goals. Milestones illustrating this change include broad government programs to reduce heart disease through prevention and control of high blood pressure and cholesterol; expansion of the U.S. Preventive Services Task Force (USPSTF) to include formerly untouchable public health problems such as domestic violence and teen pregnancy; and the establishment, however meager, of a focus within NIH to study alternative health care approaches (even prayer as a modality has received attention).

Role of Community—Expanded Concept

To fully appreciate factors affecting health and well-being, one must also take into account the community context. For the purposes of this chapter, *community* can be defined as a place shared by others in which the individual feels a sense of affiliation, of trust, of security. It also includes resources (or the lack thereof) that can enhance access to health services and influence health beliefs, attitudes, and behaviors. In the past, community has been defined primarily by geography. In today's complex, mobile society, an individual often belongs to several communities that represent various elements of life—professional, spiritual, family, professional, and so forth. It is sometimes useful to think of the individual as standing at the center of a community system represented by overlapping spheres of different sizes to indicate their importance.

However, the physical/geographic community remains important to supporting healthy lifestyles because physical environment and its health culture so directly influence the choices available. Movements toward Community-oriented Primary Care, Community Healthcare Information Networks (CHINS), and government programs such as the Healthy Communities Initiatives all testify to the recognition of a new trend for including the community orientation for delivering personal health care services. This is not just as an extension of the traditional public health arenas of preventive services and health promotion, but includes services directed toward acute and chronic illness. As I explore in more detail later, electronic mobility makes geographic considerations less important in the formation of communities, giving increasing relevance to the term *global community.*

Facing Limitations

Another indicator of the paradigm shift is the move away from sole dependence on a knowledge base kept in the heads of the medical priest-hood. For years now, health care professionals have been asked to do the impossible—store, update, and retrieve-on-demand the vast array of facts necessary to make high-quality medical decisions. Because medical care experiences most often result in successful outcomes, the public image of medical practice remains mostly positive. However, almost every family can tell you a horror story involving medical care for themselves or a relative. There is an unacceptably high error rate in today's medical practice that derives not from simple negligence or ineptitude, but from reliance on a system of training and practice that asks doctors to retain and apply a medical knowledge base that, even in narrow subspecialties, has exploded beyond anyone's capacity to optimally use or manage with traditional methods and tools.

Medicine and its practitioners have failed to squarely face their limita-tions here and, as a result, have been slow to adopt information management tools that are available and widely used within other spheres—particularly industries in which poor control of information has long been recognized to impact not only performance, but also the bottom line. Admittedly, it is the recent focus on the bottom line that is causing the health care industry to pay attention.

As health care providers, we are good at recalling the facts needed to diagnose and treat the most common problems that account for 80% to 95% of what we might see with a given patient presentation. Yet, predictably, we often fail when we encounter a problem that requires consideration of rare or uncommon diagnosis beyond our routine experience. This is especially evident in problems stemming from multifactorial causes.

The seemingly overwhelming amount of information that must be brought to bear on these uncommon or complicated problems to achieve high-quality, scientifically based diagnostic and therapeutic decisions drove us into a system of care that puts such a premium on specialization. However, specialization has provided no panacea. In spite of our efforts, patients and their problems do not fit neatly into the specialty categories that medicine has carved out. People with significant medical problems often have them against a background of health considerations that cross specialty frameworks as well as social, political, and cultural boundaries. Consider, for example, the range of specialty consultations that might be of value to a patient with heart failure (cardiologist), whose diabetes is out of control (endocrinologist), who has deteriorating skin ulcers (dermatologist/surgeon), and who is—understandably—depressed (psychiatrist). Further complicating this situation might be social, economic, and cultural factors such as poor general nutrition, poor medication compliance, or the belief that health is simply a matter of fate.

The broadening definition of *health*, the changing face of community, and the recognition of the limitations of current practice are but three of many considerations fueling this period of rapid change. And while contemplating the impact and trajectory of these changes, we would do well to bear in mind the accelerating pace of this change itself. Our society has endured changes of this magnitude before, but the pace was much different. It took at least 30 to 50 years to make the transition from an agrarian to an industrial economy. The challenge this time is to deal with changes that are glacial in scope but seismic in speed.

THE POWER SHIFT

Changes in the Professional Hierarchy

A key trend that is being accelerated by information and communications technologies is the breakdown of professional hierarchies. What was formerly held to be strictly in the domain of the specialist and subspecialist, (e.g., gastrointestinal endoscopic examinations, treadmill stress tests, etc.) has moved into the realm of the primary care physician. The primacy of the specialist and supersubspecialist has, in only a few short years, abruptly hit the skids. The Golden Age of specialization in medicine is over—at least in terms of the balance of power. Primary care has been revitalized with its presumably patient-centered values and its newly formalized institutional role as the gatekeepers to specialty services.

Can we say that the new technologies have played a role here? Perhaps, although not so much in their current incarnation as in the recognition of the inevitability of their arrival. As the powers-that-be began to reengineer the existing systems into a managed care framework, the role of specialists leapt forward as an obvious target for cost containment. It became clear that medical informatics could not only effect great cost savings on the administrative side of the business—automating paperwork and information flow—perhaps it could also reduce one of the highest cost factors, specialized clinical manpower.

Cognitive health care specialists (those that provide primarily information-oriented services such as neurologists and endocrinologists) have realized that the bulk of their contribution as direct caregivers could be made in a consultative role with the primary care provider. In this role, the human resource requirements would be greatly reduced in part because of existing supply and, as we explore later, the future use of interactive electronic media (e.g., with electronic decision-support systems and multimedia telemedical consultation). Procedure-oriented specialists (e.g., surgeons, interventional cardiologists, dermatologists) are spending more time on the cognitive parts of their specialty, for example, diagnostic referrals and follow-up such as confirming a diagnosis of gall bladder disease or following up a ligament repair. In all likelihood, much of this work could well be managed in less-expensive environments utilizing lesser trained providers supported by similar communication networks and decision-support systems—tools that potentially could magnify the productivity of a single consultative specialist many-fold.

Much of what was once deemed the professional province of physicians only, such as prescribing medications and ordering lab tests, is now being managed effectively and safely by midlevel providers such as nurse practitioners and physician assistants. Some roles formerly guarded by nurses and doctors—such as patient education for prescription medications, advice on the use of over-the-counter or non-prescription medications, splints, bandages, nutritional supplements, and even herbal remedies—have become the legitimate domain of pharmacists and even nonprofessionally trained specialists and clerks in supermarkets and health food stores. There are even franchised operations that specialize in servicing the needs of people with a particular problem—overweight, hemorrhoids, and even back pain (e.g., Relax the Back Store). Consumers who once would have gone only to their family physician, orthopedist, chiropractor, or physical therapist, can now get help from nonprofessionals selling a variety of appliances and remedies for their back pain in this unabashedly commercial environment. This is not that far from what was considered outright quackery by the professional community only a few short years ago.

Democratization of Health Care Knowledge and Tools

The dramatic changes in the health care hierarchy affect more than just health care professionals. Perhaps the most profound change is the shift of power into the hands of the consumer. Through a technology-mediated democratization of health care knowledge and tools, a suitably motivated consumer/patient may soon be in a position to be their own primary care provider.

With a better-educated populace and increasing access to health care information through the mass media, we are witnessing a shift away from the deification of our healers. Many patients now see themselves as consumers instead of supplicants, and health care professionals as educated, but fallible, mortals, providing a consumer service. And as in other areas, a larger proportion of these consumers are desiring a more active role in the decision-making and are seeking out providers that support this desire. Competitive market forces in health care are also forcing many physicians to adjust their expectations away from the professional culture that has encouraged an authoritarian approach and an emphasis on specialization. There is a new set of values emerging that encourages patient, family, and community participation in achieving better outcomes.

Furthermore, the individual health care consumer is increasingly becoming empowered to engage in fairly sophisticated self-care and health-promotion activities, with or without professional assistance, for problems formerly requiring medical intervention. Consumers can now purchase and learn to use medical devices such as blood pressure machines (syphignomanometers) and cholesterol screens, rarely seen outside of a doctor's office or hospital as little as 20 years ago. They can purchase over-the-counter medications that were some of medicine's most powerful wonder drugs only a few short years ago. Examples include nonsteroidal anti-inflammatory drugs (NSAIDS), such as ibuprofen (Advil, Nuprin) and naproxin sodium (Aleve), anti-ulcer medicines to reduce acid (H-2 blockers) such as Tagamet and Axid, and nicotine gum and patches. Lay-oriented lifestyle modification programs and resources abound as our culture encourages such things as smoking cessation, improved nutrition, and regular exercise.

Crossing generational lines, this democratization phenomenon, informed by an increasingly sophisticated and diverse medical press, has in some cases even allowed lay consumers to get out ahead of medical professionals in adopting valid therapeutic and preventive strategies. A good example of this is the use of anti-oxidant vitamins and aspirin to prevent heart disease, stroke, and cancer.

More than a few Generation Xers, the heaviest users of the Internet, rely on this vehicle as their primary source of information if they have a health problem or concern. At the other end of the age spectrum, senior citizens, long assumed to be technophobic, are becoming more active users of the internet to establish communities of common interests (Furlong, 1995). If this group, which consumes the greatest amount of health care resources, becomes more proactive and informed consumers, there could be a dramatic impact on health care systems and on the use of information technology to provide health care services.

And, of course, there are the Boomers—the most educated and consumer-aware demographic group. In general, Boomers want to be actively involved in the health care process and to have more control over health-related decisions. They are moving into a phase of their life cycle where they are becoming increasingly motivated to seek health information. Not only are they needing more personal care services for themselves, they also are assuming responsibility for family members who require the most interaction with the health care system—their aging parents and their young children.

Suffice it to say, significant changes in professional hierarchies as well as public attitudes—supported by increased information access, changing cultural norms, and new financial imperatives—are loosening the grip of professionals on the medical knowledge-base. Inevitably and irreversibly, health care knowledge and tools increasingly are becoming available to consumers.

Evolution of the Economic Model: From Laissez-Faire to Managed Care

It has become increasingly obvious that a high percentage of resources now consumed by the procedural specialties may be unnecessary and driven more by economics than science. Emerging outcome studies on regional differences in rates of surgeries and procedures continue to show a disconnect between number of services provided and real health outcomes of specialty care (Friedman & Elixhauser, 1995). These studies confirm the fact—known for at least two decades—that excess capacity in health care only increases costs and utilization—whereas quality, using outcome measures such as value (cost–benefit), evens out or falls off. As a result, the managed care industry, following the new business models of capitation (prepaid care on a one-price-per-head basis) and a bottom-line perspective, have been quick to restrict previously available but medically unnecessary services.

Once concentrated in the hands of the independent specialists and subspecialists, the power to control access and the price of subspecialty services now resides more and more in the corporate offices of managed care systems. The managed care companies are aggressively aggregating provider resources, purchasing leverage, and marketing clout. Increasingly, specialists and generalist physicians are joining these systems. Even those who only have contractual affiliations are being forced to follow guidelines that effectively constrain their decisions according to the mandates and priorities of the corporate entity. These changes are occurring in part because new information systems allow the third party (corporate entity) real-time participation in the medical decision-making loop that previously was the private domain between the individual patient and provider.

From Managing Costs to Managing Demand

In his recent review of the evolution of the managed care industry, Goldbloom (1995) suggested that the industry is now in its second stage of evolution. In the first stage, the emergence of managed care from conventional care was mostly based on price and convenience. Managed care companies were managing costs by taking the fat out of fee-for-service. In this second stage of evolution, managed care companies are being differentiated on the basis of value and quality. The companies have to compete against one another on consumer satisfaction and, eventually, on heath outcomes as consumer report cards become available (How Good Is Your Health Plan, 1996).

Perhaps more importantly, managed care companies are beginning to see that to control costs, they have to control demand for services. In fact, it is the demand-management feature of the current landscape that is driving health promotion and self-care from the back rooms to the boardrooms of the megasystems and to the forefront of health care services being offered.

In this context, interactive technologies, as a vehicle for delivering health promotion and self-care services, will make it possible for consumers to stay healthy and patients to avoid needless trips to costly facilities. New companies are springing up like weeds to offer demand-management services to the managed care industry via Internet and telecommunication links. These companies are increasingly offering new patient education and communication services including automated and nurse-assisted triage, peer support groups, and self-care tools. Inherent within these services is the transfer of health care decision making to the consumer, thereby reducing the need for provider-driven services such as unnecessary and expensive visits, tests, subspecialty care, surgery, and hospitalization.

True health promotion and prevention services on a community-wide basis await the next evolutionary stage of the industry. This is happening as megamergers and competition among managed care entities reduce the number of real players in a community. Until recently, managed care companies assumed that annually, a sizable portion of their enrollees would go elsewhere (e.g., independent practitioners, other HMOs) for care. In this environment, health promotion services, although given some lip service, carried little incentive because the investment would likely benefit a competing organization. However, in a new environment with only one, two, or three players in a community, the prepaid subscriber base begins to assume the same risk profile as the population at large. Suddenly, there is a real incentive for dealing with the real drivers of our national health care costs—the preventable illness and injury resulting from lifestyle, behavior, and the lack of healthy environments and social support systems. We are beginning to see a managed care industry that supports personal health promotion for their community of subscribers and offers an increasing variety of programs such as exercise for seniors, substance abuse prevention and treatment, nutrition for children, injury and violence prevention, and cleaning up the air and water to name a few. They may even go so far as to become involved in encouraging economic development in poor neighborhoods, development of strong families, healthy behavior, spiritual communities, and social support networks—factors that have been shown to contribute more to general health and well-being than traditional medical services alone (Berkman & Breslow, 1983; Wolinsky, 1990). In this environment, interactive technology can play an important, if not dominant, role in the development and delivery of these long-neglected health promotion programs and services.

ENABLING TECHNOLOGIES

Much as the mechanization of agriculture and industry enabled the Industrial Age, it is the new information power tools that will provide much of the leverage as we move from an industrial to information age. As consumers begin to assume a greater role in their health care decisions, they will need (even demand) better tools to access health information. Here is where the new information and communications technologies will play the biggest role in transforming the balance of power.

A few specific information and communications technologies deserve special mention here because they will play key roles in facilitating the health care power shift. As I describe these, keep in mind that considering them in isolation is a bit like describing the human organism from the perspective of an individual medical specialty or basic science discipline. Individual technologies should be examined in the broader context of how they will likely converge—such that the whole is more than the sum of its parts— into specialized electronic performance support systems (EPSSs) that connect all the players in the health care equation to optimize and facilitate health-related decision making and activities. In principal, an EPSS is an information power tool that enables end users to perform certain activities (traditionally job-related) at a significantly higher level than their formal training or experience would otherwise allow. These systems allow users to identify, access, integrate, and present just-in-time information that will facilitate decision making or psychomotor performance. EPSSs have already made dramatic differences in the re-engineering efforts of companies ranging from financial services to the automotive industry, and they will likely become widespread in health care—further altering the roles, responsibilities, and relationships between the individual consumer, health care professionals, and health care institutions.

The Microcomputer and Warp Speed

We would do well to bear in mind that many of these technologies are following the path of the key technology breakthrough that started it all—the personal computer. This little machine arrived on the scene a little more than two decades ago, yet it has profoundly touched almost every aspect of human endeavor—including virtually every step of creating and publishing this book. Moreover, if it were not for this machine and its implications for your life, you would not even be reading this book.

Just how mind-bogglingly quickly this tool is evolving is underscored by the fact that most mortals have to resort to analogy just to try to comprehend it. A favorite analogy of technopundits measures microcomputer evolution against that of the most familiar and revolutionary technology of the Industrial Age, the automobile: If cars were on the same price/performance curve as computers (Moore's Law states that microprocessors get twice as powerful and cost half as much every 18 months), a Rolls that cost $100,000 in 1970 would now cost $.80, get 1,000 miles to the gallon, and would travel at mach 6! Hold onto your chronobiologic hat!

Computer-Based Patient Training (CBPT)

Over the past three decades, computer-based training (CBT), often referred to as computer-based instruction (CBI), has shown itself to provide superior effectiveness and efficiencies to traditional teacher-led approaches in many settings. First clearly established in military training, many industries requiring highly trained technicians have adopted the medium to great effect. Now we see widespread acceptance of CBT in public education, especially in math and science where automated presentation and use of powerful simulations augment understaffed and underequipped class-rooms.

The use of computer-based teaching in health care is likewise gaining sway as the value of automated, individualized, self-paced learning becomes more evident in the instruction of both patients and providers. Research has confirmed that CBT can lead to increased knowledge and behavior-change that has far-reaching implications especially in areas such as medication compliance and self-care in chronic diseases like diabetes, asthma, AIDS, and arthritis (for a review of these studies, see Kahn, 1993). Many of these studies also have established that there is a high level of patient acceptance for interactive technology used in this way. Recent implementations that focus on patient decision making have even suggested large, potential-cost impact in some areas such as prostate surgery (Hayward & Kahn, 1995; Kahn, 1993). To date, most of these CBT programs have been available in health care facilities only as stand-alone applications. In the near future, however, we are likely to see these programs become more widely available and distributed over ubiquitous networks (such as the Internet), providing communication links to providers along with auto-mated decision support.

Digital Documentation—Medical Records Go Electronic

Moving to a more central health care information arena, perhaps one of the most important mediators of change will be the advent of the computerized patient record. Our medical records are considered the primary tool for communication and storage of information relating to our health; they are supposed to document, organize, and codify all the complex data gathered by health care providers. This information should then be available when-ever and wherever it is needed to provide the basis for high-quality, data-driven health care decisions.

Unfortunately, most medical record systems today do not achieve this standard. As often as not, they are incomplete, incomprehensible, or simply unavailable. They often contain a hodgepodge of data that lack coherence. Medical record systems today usually fail to truly represent the individual's health status and needs; with few exceptions, they rarely try to systematically address the full spectrum of health (including emotional, spiritual, and economic factors) that affect well-being; they often neglect, as does the present mainstream delivery system, the importance of prevention and health promotion. They can also be misleading because they represent the bias of the provider that inscribed or dictated them; they usually fail to incorporate the individual patient's perspective in the information loop—information needed to enable the patient's meaningful participation in the decision-making process.

However, driven by the current wave of re-engineering emanating from the aggregation of institutions and practices into mega managed-care systems, there is a broad consensus that a single electronic patient record, driven by both administrative and clinical considerations, will provide a much-needed infrastructure for the health care of the future. There also seems to be a spreading awareness that allowing patients to access and update their medical records—supported by appropriate technology—can be used to much greater advantage in facilitating cost-effective health care decisions. In fact, most large managed care organizations are considering or are already experimenting with home-based computers to provide patient reminders, leverage their triage functions, facilitate appropriate self-care, as well as to gather previsit patient data.

A specific example of an evolving consumer-centered medical record system is under development through a joint venture between Netscape, a company that provides Internet access to the World Wide Web, and Health Desk, a California-based company. They will provide individuals with their own medical record that they keep, update, and use to communicate with their health care provider(s) over the Internet. This software product also will include a graphics-based self-care section.

Similar products will no doubt be made available through other major consumer health information services currently under development at established online providers like America Online and Prodigy as well as major new players including IBM, AT&T, Time-Warner, and Microsoft. If done well, these could be of great value both to individuals in making their own health decisions and to managed-care companies in their efforts to control the use of facility-based resources.

Telemedicine—Getting Care Without Going There

Another major technology front advancing at an unprecedented rate is telecommunications. The concept of telemedicine describes the combining of telecommunications and information technologies to deliver medical services. Negroponte (1995), the MIT guru of the new media, has suggested the technology now gives us the power to move electrons instead of atoms. As we reengineer information systems, a key question will be how to deliver health care services without having to transport the patient and without requiring bricks and mortar. Telecommunication channels have the potential to greatly enhance the power of consumers to access health care resources including information for self-care and services offered by managed care providers.

The same technologies that are allowing knowledge workers everywhere to telecommute can, when augmented by a few additional sensors such as an electronic stethoscope, enable physicians and other providers to make electronic housecalls. Even surgery may soon be performed remotely by using a suite of incredible new technologies that create a sense of telepresence. Gloves with sensors that can provide very fine granularity to control remote robotic hands while providing both real-time tactile and visual feedback have been successfully demonstrated with actual patients (Bowersox et al., 1996; Kaplan, Hunter, Durlach, Schidek, & Rattner, 1995). Potentially, this would allow patients anywhere in the world—given local availability of the technology—to receive the services from the most capable providers they could find even if they are at a geographically remote center of excellence.

The rate at which telemedicine will play a role in the paradigm shift will depend on several factors. One of these is the bandwidth issue, often spoken of as the size of the pipe in referring to the capacity of the wiring or wireless technology to carry the needed information signal. In particular, it is assumed that in order to provide the user-friendly interfaces needed by consumers to conduct the transactions over these new media, much higher bandwidths than currently exist will be necessary. However, at the present rate of evolution, this is likely not to be a limiting factor for long. What is a more important challenge is getting an appropriate health care-oriented terminal to the point-of-need such as in the home or office. This will be an issue long after the bandwidth problem is addressed.

In addition, more serious bottlenecks likely will occur in redesigning (re-engineering) and implementing the administrative and services functions from the current paradigm of health care delivery to new models of delivery required in a digital age. It took well over a decade for the industry to embrace outpatient surgery—even in the face of compelling cost and outcome data. Similarly, health care organizations will find it difficult to make the shift from facility-based, provider-centered care to technology-mediated clinical tele-encounters. Even in this climate of rapid change, we can expect to see a good deal of institutional inertia; this is inherent in such an established and complex system, especially when the stakes are so high.

Before telemedicine can become a reality, a host of other questions remain to be answered as well; to wit: Who has professional licensure jurisdiction when patient and provider are in different states? Is the patient coming to the physician or vice versa? What and how much of the healing is lost without the physical presence of a face-to-face encounter? What can be done by remote control and what resources are needed on both sides to control safety and efficacy? Experiments dating back to the sixties have established the safety and efficacy of teleclinics (for a review see Crump & Pfeil, 1995; Perednia & Allen, 1995). New technologies are simply adding orders of magnitude to the fidelity in simulating face-to-face clinical encounters.

Health Care Intelligent Agents

It may be that even before provider systems fully make the transition to telemedicine, we will begin to cross technology thresholds that enable consumers to become more proactive participants in their own health care decisions. New mass distribution communications technologies are arriving on the scene, such as cable modems, set-top network computers, and direct satellite TV/POTS (plain old telephone systems) hybrids. These will quite possibly overcome the problem of consumer access, at least for most people.

A disturbing aspect of the business models being suggested for these technologies is that they do not adequately address how we will extend the reach of these systems to the poor and the information-disenfranchised populations residing in undeserved areas. Even for the majority, another problem may emerge: the problem of information overload. A general search on the Internet can now result in an experience that has been characterized as filling a water glass with a fire hose. The large quantity of information available on a given topic often lacks sufficient quality in that it is incomplete, outdated, inappropriate, and often provided by someone with a commercial agenda and not necessarily in alignment with the best interests of the patient/consumer. Thus, the information seeker needs a filter to help put the information into a usable form for making health care decisions. This implies not only selecting and weighing available quantifiable data, but applying judgment and wisdom—and this is where the professional provider role remains most important.

One approach to dealing with the overload issue is the use of intelligent agents, a form of technology also labeled Knowbots. Conceptually, one can think of these agents as functioning as a personal electronic health care advocate. Residing in your personal electronic filing space (e.g., a desktop computer) as a special, self-contained program, your health care intelligent agent (HIA) would actually know and continuously learn all the facts about your health and family history, preferences, and health status. Whenever a health problem arises, the Agent would then be able to go into cyberspace on your behalf to retrieve all relevant and specific information at the time of need, filter, and then present it in a usable form. Taken to its logical extreme, this agent might even negotiate with your provider system's electronic agents—with or without human intervention—and bring back the specific advice or prescription to fit the situation. This is not meant to imply that these machines will replace all professional encounters; these agents are simply programs that serve as a means to seek/manage/filter health information, customized by the patient/consumer to help meet his or her health needs. Similar tools will also be used by the professional provider—who also can suffer from this information toxicity.

Direct Consumer Access to Tools of Medical Technology

As the new health care economics drives the concept that the consumer and the family can and should be the real primary caregivers, we will see more testing and data-gathering technology oriented in this direction. We can expect to see an array of physiologic monitoring devices (e.g., blood pressure cuffs, glucose monitors, peak flow meters) becoming a common feature in both residential and workplace environments, enabling patients and available caregivers to gather basic clinical data such as temperature, pulse, blood pressure, respiration, heart and respiratory sounds, and even EKG and EMG signals. Laboratory tests recently announced include home tests for cholesterol and AIDS screening, along with the widely used home pregnancy tests. These devices and tests, formerly available only through expensive provider-based facilities, can be combined with simple communications systems to transmit data to health care providers. Sometimes the results will be immediately available and transmitted for decision making; sometimes it will require the use of new, but easily accomplished, transportation/courier paradigm, to make it all happen—but happen it will. A video camera and microphone opens the door wider still. These will evolve from portable tools used by home-care itinerant providers to integral components of fully integrated, easily accessed health information and support systems that I predict will become standard features of the health care landscape—be they located in our homes, neighborhoods, workplaces, or other community gathering places.

Networks—The Internet, World Wide Web, and Beyond

Another key technology here, and perhaps the one most visible in current discussions of the power shifts, is ubiquitous networking such as that provided by the Internet. Many individuals are already using online services as a primary source of health information and support. This is especially true in areas of chronic diseases and other emotionally taxing problems—AIDS, cancer, rare (e.g., childhood genetic disorders) and controversial diagnoses (e.g., chronic fatigue syndrome, PMS)—where the formal health care system is ill-equipped to serve the legitimate needs of these people, especially the social and emotional support needs of patients and their families.

In his new book, Furguson (1996) described the world of online support groups, giving poignant examples of how the new interactive technologies are both supplementing the traditional delivery system and forging new medical information architectures and community subcultures. Although there is certainly misinformation in the more than 10,000 health-related sites available online, anyone can now search the indexed medical literature online via Medline and access high-quality, lay-oriented resources in just about any area they might imagine. Furthermore, ubiquitous interactive networking of digital multimedia information will enable the linkage of geographically independent health care organizations that will allow them and their patients to effectively integrate health care resources and services.

THE PATH: ONE SCENARIO

To offer one vision of the health care future, consider how these and other key information and communications technologies could be synthesized into electronic performance support systems specialized for healthcare applications that might be called Health Action Performance Information Support Systems or HAPISS (pronounced "happies"). Initially, the elements of these systems are becoming visible as stand-alone or networked programs, offering a limited amount of health information, communication, and decision-support resources (see, for example, chapters 5 through 10 in this volume). As they evolve, however, we will see even higher levels of integration both of technology—informatics and telematics—as well as health care content. These systems will become increasingly user-friendly, comprehensive in scope, and organic in nature. These HAPISS will be used by all players to perform their new roles separately and in connection with each other, transforming the relationships between consumers and providers. Ultimately, they will become transparent extensions of individual sensations and psyches providing a seamless bridge between the domain of provider and the needs of the consumer.

What can we expect in the near term? We are in a period where we will see increasing de-institutionalization of care. HAPISS will allow more care to be moved to the point and time of need, such as into the home, neighborhoods, and the workplace. HAPISS will change the manpower equation by allowing new types of caregivers with different skill sets to replace traditional types and hierarchies of professionals. In many situations in the future, a technician or paraprofessional armed with HAPISS will be able do the job of a highly trained professional— a professional who now requires lengthy and expensive training to acquire the knowledge base and skill set necessary to provide technically competent care. New types of providers and new ways of training them will emerge to fit the new delivery paradigms. The training and practice model where the physician/scientist is the primary resource is likely to lose currency. The focus will be on what type of professional, paraprofessional, or lay extender is required to provide cost-effective, competent, and compassionate care. In this environment, there will be new priorities in training and practice. A high premium will be placed on interpersonal skills, on information tool utilization, as well as technical psychomotor skills.

I project the emergence of two new tiers of professionals. This first would be a personal health care advocate (PHA) who will serve as the primary human interface between the patient/consumer and the rest of the delivery system. This advocate/coach will rely on interpersonal skills and a trust-based relationship to help the patient/consumer focus their decision-making—often using an EPSS—which includes direct links to appropriate provider expertise.

The second type of professional is a new breed of health care technician. These technicians will be prepared to enter the workforce with sharply focused, competency-based training. They will stand on the front lines, supported by EPSSs, to perform most routine technical services including invasive procedures such as biopsies and angioplasties, as well as many other routine surgeries.

Let's take a look at a specific example: Who would you rather have perform your knee surgery? A professor of orthopedics who spends 75% of his time in research and administration; an orthopedic practitioner with good training and good hands, 3 to 4 years in practice and who has performed 200 cases of knee surgery like yours or; a knee surgery technician (KST) using an EPSS. This KST is selected into training based on manual dexterity, mechanical aptitude, emotional maturity, and interpersonal skills. She also has 3 to 4 years experience; has performed 3,000 cases like yours; spends 80% to 100% of her time doing this type of surgery; is continuously monitored for process and outcome (she's given quality improvement feedback for every case); has real-time intraoperative access to expert systems (for identifying and solving unusual problems) based on a database of up-to-the-minute, world-wide experience; and, if necessary, can access through the EPSS, a panel of sub-specialist consulting knee surgeons anywhere in the world. These consultants can both view and even participate in your procedure (using telepresence technology). Follow-up care would be delivered by a rehabilitation technician (could be same person) with similar technical background credentials and access to a knee rehabilitation performance-support system.

The technician-based approach offers the advantages of supersubspecialization (e.g., an orthopedist with practice limited to knee surgery)—including safety and efficacy—but at a much lower cost by housing the scientific knowledge-base in the EPSS and utilizing only the resources and that portion of the skill set required to optimize outcome.

On the other side of the equation, a HAPISS will enable patient/consumers to comfortably manage many common problems now requiring professional attention using only self-care measures supplemented by inexpensive technical support (procedural technicians) and, when necessary, by a coach/advocate. This means the consumer/patient can assume a much greater role in health decisions.

As an example, suppose you sprained your knee skiing. How would you access your own HAPISS in this situation? You limp in from the ski slopes, sit down before the display in your hotel room, prop up your knee, and start your automated clinical encounter by inserting a smartcard in the reader on the set. The card contains your personal Healthcare Intelligence Agent (HIA), the special self-contained program that knows or is programmed to access the database containing your health records—all the facts about your health and family history, preferences, and health status. You state your name (your voice print serves as your password). This system now has access to your medical records, your preferences, and programmed communication links to your designated providers. The HIA questions you regarding your symptoms—"How did it happen? Where and when does it hurts? Did you hear a pop?" It generates a hypothesis based on this and concludes that it likely is a minor to moderate sprain, provides interactive information and advice (written and audio–visual) about using ice, immobiliazation, anti-inflammatory medication, and rest. It might also advise on how to avoid further injury now and in the future and under what circumstances to check back in with the HAPISS or go to an emergency facility. Your record is updated accordingly. Your PHA (from your neighborhood back home) is notified. He or she appears on the screen at the end of the session or sends an electronic message shortly thereafter to ask how you are doing and reinforce the instructions.

Now let us say you followed the instructions to use ice, elevation and anti-inflammatory pain medications (Ibuprofen) for 2 days, along with a recyclable knee imobilizer (delivered that day by the health care courier service contracted to your managed care organization). A week later as you get ready to leave for work (you have a ride with a coworker), your PHA calls you to your HAPISS, links you to a knee injury consultation program, and the two of you—on a teleconference or, if practical, in person—go through a simple procedure to remove the immobilizing brace and safely perform a self-examination on the knee. You determine that the pattern of your pain and the potential instability you detect on the interactive video-guided self-exam, visually monitored by your PHA, could suggest something more than a simple sprain. If physically present, your PHA could perform an expanded exam. As your PHA enters your response, a orthopedic paraprofessional, automatically called by the logic programmed into the system, joins the teleconference to guide a further confirmatory self-exam before scheduling a mobile imaging van to come by your worksite during your afternoon break. At that time, a diagnostic technician does a confirmatory physical exam and performs an imaging study analogous to today's sonograms and MRI—but, of course, this is done with a relatively inexpensive, hand-held device. This study, completed in a matter of minutes, confirms through a pattern-recognition program keyed to a knee injury database, or if necessary, in consultation with a remote imaging specialist (teleradiologist), that you have torn cartilage. It offers a precise characterization of the injury and the arthroscopic procedure required to correct it.

Back at home, as you sit in front of the wall-hung display of your HAPISS terminal, you receive a detailed, interactive, patient-education session, customized for you based on your medical records, your learning style, cognitive abilities, cultural biases and preferences, and provider specifications. Through this, you learn all about your problem and about a certified knee technician surgery program. You get all your questions answered through the interactive system, including accepting the offer to be automatically connected to two individuals like yourself who have recently undergone the procedure. You now give your truly informed consent (voiceprint confirmed) to undergo the arthroscopic procedure and rehabilitation required to optimally correct the injury and minimize pain, discomfort, and lost time.

Going back one step further, let us look at how you would have picked your PHA and your current health care provider system. When you got your new job, your employer gave you an electronic voucher that represented financial participation in a capitated plan. You sat down at your HAPISS in your living room and launched a query to go shopping for your new healthcare provider system. Your HIA asks you a series of questions regarding your preferences and feelings about past experiences (it knows your priorities based on family and medical history, as well as any cultural and social sensitivities), displays rankings of available service providers based on your preferences, and current quality indexes (comparative outcome and satisfaction measures). Your HIA electronically negotiates with the top choices of systems and comes back with video biography of potential primary providers—you are looking to choose your primary care physician and a Certified Healthcare Advocate (CHA). From these, you select and interview (via teleconference or in person) your top candidates and some of their patients, electronically tour their facilities, and then make your choice.

Systems like the hypothetical HAPISS also are referred to with a variety of other names such as CHISs (Consumer Health Information Systems) or PHIS (Personal Health Information Systems). These systems are beginning to be developed not only by managed-care organizations, but also by large communication companies who sometimes team up with health care organizations. In fact, this may be shaping up as the real competitive battleground for health care industry in the decade to come—a brave new world where interactive digital technology will play a major role in shifting the balance of power from providers to consumers.

CONCLUSION

In order to relate the new paradigms of the Information/Communications Age, it seems that journalists have attached the prefix *Cyber-* to half the words in English language! This term derives from the Greek word for steersman or navigator. *Cyberhealth*, in the new millennia, will mean that the consumer will transition from the role of passenger (most often taking a back seat) into the driver's seat—making health care decisions supported by a specialized, health-oriented EPSS (e.g., a HAPISS) that provides information, communications, and cyberlinks to a host of new, efficient health services and providers. And with the consumer in the driver's seat, there will be a new emphasis on "defensive driving"—for example prevention-oriented health behavior. Here, too, the individual will be encouraged and supported by the HAPISS—at the time of need—providing useful information (knowledge) and links to a community of peers, lay and professional coaches, advisors, and consultants—all focused on correcting, maintaining, and promoting optimum health in the broad World Health Organization definition of the word.

REFERENCES

Bowersox, J. C., Shah, A., Jensen, J., Hill, J., Cordts, P. R., & Green, P. S. (1996). Vascular applications of telepresence surgery: Initial feasibility and studies in swine. *Journal of Vascular Surgery, 23*, 281–287.

Berkman, L. F., & Breslow, L. (1983). *Health and ways of living: The Alameda County Study.* New York: Oxford University Press.

Crump, W. J., & Pfeil, T. (1995). A telemedicine primer: An introduction to the technology and an overview of the field. *Archives of Family Medicine, 4*, 796–803.

Friedman, B., & Elixhauser, A. (1995). The changing distribution of major surgical procedure across hospitals: Were supply shifts and disequilibrium important? *Health Economics, 4*, 301–314.

Furguson, T. (1996). *Health online: How to find health infomration, support forums, and self-help communities in cyberspace.* Reading, MA: Addison-Wesley.

Furlong, M. (1995). Communities for seniors in cyberspace. *Ageing International, 22*, 31–33.

Goldbloom, J. (1995) Managed care comes of age. *Health Care Forum, 38*, 14–24.

Hayward, R., & Kahn, G. (1995). Patient education. In J. Osheroff (Ed.), *Computers in clinical practice.* Philadelphia: American College of Physicians.

How good is your health plan? *Consumer Reports, 61*(8), 28–42.

Kahn G. (1993). Computer-based patient education: A progress report. *M.D. Computing, 10*, 93–99.

Kaplan, K., Hunter, I., Durlach, N. I., Schidek, D. L., & Rattner, D. (1995). A virtual environment for a surgical room of the future. In R. M. Satava, K. Morgan, H. B. Zieburg, R. Mattheus, & J. P. Christsen (Eds.), *Interactive technology and the new paradigm for health care* (pp. 161–167). Burke, VA: IOS Press.

Negroponte, N. (1995). *Being digital.* New York: Knopf.

Perednia, C. A., & Allen, A. (1995). Telemedicine technology and clinical applications. *Journal of the American Medical Association, 273*, 483–488.

Wolinsky, F. D. (1990). *Health and health behavior among elderly Americans: An age-stratification perspective.* New York: Springer.

12

Facilitating the Adoption of New Technology for Health Promotion in Health Care Organizations

William R. Gold, Jr.
Blue Cross Blue Shield Minnesota

New advances in computing technology offer tremendous opportunities for health promotion, patient education, and the delivery of health services. A growing body of research suggests these programs can be very effective if carefully developed and implemented. The implementation phase may be as challenging as the development phase. A major obstacle to implementation appears to be the reluctance of health care organizations to invest in and use the technology. The purpose of this chapter is to discuss characteristics of organizational environments that account for this reluctance and to suggest ways that program developers and organizational leadership can respond to these challenges.

BACKGROUND

As one considers newer applications for computer technology, especially as it relates to patient education and outcomes, the backdrop against which the technology will be introduced must be considered. Leaders in health care have always faced the challenge of balancing the mission of their organizations with insuring the solvency (and profitability) of the business. Although change has been a continuous reality in health care, at no other time has change so dramatically impacted what has become to be called the *health care industry*. In addition to understanding national trends affecting the industry, it is also useful to appreciate the internal challenges that organizations face during these times of change. It is the organizational level that determines the role technology will play in the health care equation. Considering these unique organizational characteristics may help to develop a more effective implementation strategy.

Since 1980, there has been an ever-increasing realization that medical care in the United States costs too much. Whereas there is disagreement on the cause of the rising cost of care, there is little disagreement about the need to reduce the cost. Cost containment has led to the emergence of managed care, perhaps the most pervasive trend in health care today.

MANAGED CARE

Although there are many definitions of managed care, its central themes are: shifting the risk, decreasing the cost of care, increased documentation of quality and value, and care for a defined population. It is useful to understand how managed care differs from the traditional mechanism of funding and management of care.

With traditional indemnity health insurance coverage, the risk for rising health care costs is born by the payer, frequently the employer. The providers (doctor, hospitals, and others) set the fees for services and the insurance company calculates its premiums and passes the costs to the payers. As medical care costs rise, the premiums rise. The risk for the cost of care is born primarily by employers. As a contrast, in advanced managed care markets, the risk is shifted from the payer (employer) to the providers (doctors and hospitals). Under this payment mechanism, the providers are given a fixed amount of money (capitation) for each person covered. All the care is provided with this capitation amount. The providers are now at risk for any gap between the amount of money available and the cost for that care. They are at risk for waste, duplication, and unnecessary services. Care now must be managed to reduce unnecessary services, inefficient care, and to eliminate processes that do not add value. The administration of this mechanism of payment and management of care is frequently done by a health plan (health management organization; HMO) that may also assume part of the risk.

Managed care has grown dramatically in the past decade. It is estimated that 95 million people in the United States are members of an HMO through which health care is financed on a prepaid basis (Rivo, Mays, Ketzoff, & Kindig, 1995). This is triple the number of members just 10 years ago.

MEDICINE AS A BUSINESS

A separate but related trend in medicine is the corporatization of medicine (Relman, 1980). Medical care in the 1970s and early 1980s provided the ability for some components of the industry to reap substantial profits. Because the risk of medical costs prior to managed care were that of the payer, it was virtually guaranteed that a profit could be realized. Medicine was looking more like a business and less like a profession with the appearance of for-profit hospitals, free-standing investor owned facilities such as ambulatory surgery units or radiology facilities, and increasing data documenting overuse of certain medical services presumably because of payment incentives. HMOs in some markets realized huge profits as efficiencies in how care was delivered improved. It was recognized that profit could be increased as the volume of covered lives increased.

As medicine became more like a business, it was subjected to the familiar market forces of supply and demand. In certain regions of the United States, the efforts of health care systems to become more efficient sometimes exposed an excess capacity. That is, too many hospital beds and too many doctors. The reaction to these realizations was consolidation and competition.

Over the last several years, there have been many signs of consolidation and competition with the growth of managed care. Free-standing hospitals were partnering with other hospitals in various business arrangements to respond to the pressures of cost containment and competition. Physicians were forming networks as a method to band together to insure that they remained players in a changing market. New business arrangements were developed. For example, hospitals and physicians were organized as physician–hospital organizations (PHOs). The appearance of large for-profit hospital chains became a new phenomenon in health care. The rapid growth of some of these hospital systems led some to predict in the mid-1980s that the United States would end up with two or three health care systems, each competing with the others to grow even larger.

Although the prediction about the number of systems may have been inaccurate, the consolidation movement continues. Most consolidation is toward the integrated health care system where most of the medical services (outpatient and inpatient care) are provided by the same corporate entity. This business configuration is thought to have an advantage for competing for managed care lives because of its ability to provide coordinated care, collect and analyze data, and manage access and service to members.

POPULATION BASED HEALTH CARE

The originators of managed care argued that capitation would help shift the health care system from a piecework mentality to one based on care of a population. The health of the population would be the responsibility of the providers, who would consider both short-term and long-term needs of the patient. Keeping the patient healthy now becomes a strategic goal that was never fully embraced under the traditional, fee-for-service sickness model of care. Population-based care leads to the need for better measurement of health status (outcomes), new preventative strategies, and defining a new role for the patient as a partner in maintaining health. Smoking cessation, exercise programs, extensive nutritional support, infant car seat giveaways, and bicycle helmet safety programs are but a few programs designed to involve the patient in prevention efforts to maintain or improve health status. Through these strategies, the total cost of care becomes lower.

INCREASING ACCOUNTABILITY

One related trend resulting from these changes is the increased interest in documenting and improving quality. When the for-profit (proprietary) hospitals re-emerged on the landscape of American medicine, the balance between quality versus profit was questioned. As a result, there exist new demands for accountability. Measuring dimensions of quality has become a major activity in order to document that the care provided was of high quality and appropriate. Although some view this as an intrusion and unnecessarily burdensome expense, there are some advantages for this new level of accountability.

Health maintenance organizations now are routinely required to supply data that indicates how well their provider network is performing. Patient satisfaction is one dimension of these new measures. Furthermore, there is a request by the payers for more data about prevention services. These trends in medicine on a macro level have obviously had a profound effect at the local level. Understanding how individual organizations confront these global changes gives one the basis to develop an effective strategy for introducing new technologies into the organization.

Many health care organizations have maintained a reactive mode to many of the changes described. They are struggling simply to survive and find their niche in the market. Others have been proactive as the future became clearer to them. They have exercised creativity, strategic thinking, and have implemented innovative changes to improve the health of the population they serve. These different organizational climates can be either resources or barriers with respect to the adoption of interactive technology for patient education and other aspects of health promotion.

ISSUES WITH A POTENTIALLY NEGATIVE IMPACT

Short Term Thinking

Many organizations are finding themselves in an increasingly competitive environment in which most activity must be directed toward short-term gains in cost containment or market share. The payback period for investing in improved prevention services and better informed patients is compared with the need to reduce utilization expenses next month. Organizations embracing short-term thinking are not likely to divert capital to new technology that does not produce immediate results. Identifying specific applications with short-term payback may be one strategy to employ with these organizations. For example, patient education around appropriate use of the Emergency Department or outpatient clinic may be viewed differently than an application that increases compliance with a low-fat diet. Because the focus on short-term thinking exists in certain health care markets, new technologies will be adopted only to the extent they can produce quick results.

The Burden of Change

Organizations are not static. They are more likened to a living organism—changing and growing as time passes (Hindle & Lawrence, 1993). Some of the changes are planned and deliberate although many are unplanned. Changes in organizations consume energy and resources, a fact vastly underappreciated. The cost in dollars is usually not even accounted for, but astute leaders are aware of it. Whether measured or not, these costs may be called the burden of change. For some organizations, this burden may be so great that, even if there are clear advantages to new technologies for their systems, the additional burden of creating further change to what exists precludes taking on any new projects. The timing is wrong. As one develops a picture of the organization, it is useful to determine what major changes are underway, what significant changes have already been planned, and how new technologies fit within these changes.

Competition With Other Technologies

Innovations in information technology are impacting several processes within organizations and prioritizing these innovations in a meaningful way has become very challenging. As one considers a strategy to introduce a new technology for patient education and health promotion, there already may be significant demands to invest dollars in new technology for other purposes. For example, the need to develop an electronic medical record may be perceived to be more important than enhancing patient education capabilities, particularly if patient education and health promotion are seen as separate from the medical record (for an analysis of how these can be integrated, see chapter 11, this volume).

Considering the existing financial commitments to other technology programs, it may be necessary to start small with one demonstration project that might clearly document the effectiveness of interactive technology with the intent of expanding the program later.

Decision-Making Capabilities

Each organization has a set of unwritten rules that determine the behaviors of the people who work in the organization. This has been called the *culture of the organization* (Luthans, 1995). One characteristic of organizational culture is how decisions are made and carried out. Answers to questions like "Who makes the final decision?"; "How long does it take for a decision to spend money?" and, "Is there accountability for carrying out the plan once a decision is made?" will impact directly on the ability to successfully introduce technology into an organization.

For example, although some hospitals have a physical space labeled "Patient Education," the majority of patient education takes place in the physician's office or on the hospital nursing unit. As a result, in most organizations, patient education activity is a decentralized process whereas budgetary decision making is very likely centralized. A request from one department to spend $25,000 for computer technology to improve patient education for patients with a particular disease may seem perfectly rational to the requesters. They may even have made a plausible case for a substantial return on investment as part of their argument. Although the decision makers in the administration may agree with the request, they may see another problem on the horizon. This may be the first of many requests for computer technology throughout their organization. The request for $25,000 may mushroom to $500,000 as the need for new technology spreads in the organization. This becomes a major dilemma for the organization. It reminds one of a parent who, having said "yes" to one child, is now caught having to say "yes" to all the children.

Given this scenario, a delay in the decision is likely, and an unfavorable decision may also result. One strategy that may impact this situation would be to involve the administration early in the process and help them develop a framework to measure the effectiveness of the proposed new technology. Collecting outcomes data associated with the new technology will allow administrators to make more rational decisions as others in the organization also develop proposals for introducing new technologies into health care.

ISSUES WITH A POTENTIALLY POSITIVE IMPACT

As integrated health care systems mature, their mission may become clearer and a change in thinking may take place around certain themes. These themes are overlapping and synergistic. Understanding which of these themes is at work in a particular organization will help developers introduce their new technologies in the most favorable light.

Planning as a Theme

As systems grow larger, their use of planning as a business strategy may increase. Some health care systems have well-defined plans built from well-conceived mission and vision statements. These organizations are internally aligned and know the direction they are headed. There is a sense of rationality about what they are doing to meet the challenges of the new order of health care. Although there are many organizations moving toward this configuration, both philosophically and structurally, there are countless more that have only begun to consider how best to adapt to the challenges of managed care. They may still be distracted with the day-to-day realities of massive change. Organizations who have employed planning as part of their strategy have probably altered their thinking about how to care for a defined population. A number of related themes may also be in play, such as innovative methods for delivering health information to their members. Each of these shifts in thinking may be important adjuncts for introducing new technologies into an organization.

Health of a Defined Population and the Total Cost of Care

The integrated health care system, which has as its mission the health of a defined population, begins to redefine the traditional sickness model of care to something quite different. In the sickness model, patients determine when they are sick and the organization (clinic or hospital) responds. The organization's resources are consumed largely as a reaction to the demands placed on it by the patient population. Under the new paradigm, health care systems are assuming the responsibility for maintaining and improving the health of a defined population and also the risk for the costs associated with this activity. The organization now thinks about how to reduce the burden of illness and, therefore, the total cost of care. This transition is clearly evident in the area of preventative services. Most organizations have not assumed a proactive role in assuring full compliance with recommended preventative services. Under this new paradigm, strategies in providing proven prevention services are aggressively pursued, such as childhood immunizations or mammography screening.

Changing health behaviors that drive up a substantial portion of the health care costs is another responsibility that organizations are ready to consider. For example, organizations may now see a new role for themselves in reducing the incidence of smoking or obesity in their populations. Likewise, there is increased interest in modifying behaviors in patients with chronic diseases. This activity may come under a number of rubrics such as case management, demand management, or disease management. The common thread in these programs is decreased utilization of health care services through patient education, ongoing interaction with the patient, improved coordination of care, increased patient compliance, and better patient decision making. The foundation of these programs resides in better-informed and involved patients, and earlier recognition of recurrence and the need for intervention. This represents an excellent platform to introduce advanced patient education and health promotion technology.

Awareness of New Technologies

Many large organizations have sophisticated systems to track utilization and monitor clinical quality improvement. Now, with more affordable and powerful computing systems available, even smaller health care providers have (or have access to) sophisticated communication and information technologies. Experience with computers is no longer something reserved for a few technical folks in the business office. Tapping into this computer expertise within these organizations is an important strategy for increasing organizational acceptance of new technologies.

The Patient as a Partner

As newer models of health care are considered, the role of the patient in the care process is also changing. The view that the patient is a passive player in the health care system is fast becoming outmoded. Although the patient's role in utilizing health care has been long recognized, only recently has the patient been considered a key component in an overall strategy in managing those services. Incentive programs to induce patients toward educational programs and ongoing care are increasingly being used to further involve patients in the equation of how to increase quality and decrease costs.

The changing role of the patient is being explored on a number of levels. Twenty-four-hour nurse-on-call phone systems allow patients to ask a friendly voice what they should do. Should they bring their child to the emergency room at 2 a.m. because of a fever? Also, the Internet is exponentially increasing access to health information. Although the accuracy and effectiveness of this information is unproven, it is a modality that can be harnessed and used effectively in helping people make decisions about health and medical care. Furthermore, interactive, patient decision-making programs have been established for a number of diseases (e.g., benign prostate hyperplasia, breast cancer; Barry, Fowler, Mulley, Henderson, & Wennberg, 1995; Street, Voigt, Geyer, Manning, & Swanson, 1995). In addition to helping patients better understand treatment issues, these programs also try to bring the patient's own values into consideration of a medical decision. If used prior to a visit to decide treatment, these interactive programs may effectively increase patient involvement in the decision-making process that, in turn, could lead to better outcomes (see, for example, Kaplan, Greenfield, & Ware, 1989).

Accountability and Determination of Value

Another development that could have a significant influence on how newer technologies might be introduced is the increased interest by the payers of health care for providers to demonstrate the value of the care provided. By no means a mature science, the trend is clear. There are increasing requirements to develop, collect, analyze, and improve on measures of quality and effectiveness (outcomes).

Historically, documenting the quality of patient care and services has never been a central activity of most health care institutions. However, with the growth of managed care and the ability to look at costs across geographic sites, there is an increasing demand for more and better measurements of quality. In the past, measures of quality were often done because they were required, usually as part of an accreditation process. Now, however, outcome measures are being used to make significant improvements in the process of care. Gaps between performance and what's possible are being viewed as opportunities and not as problems.

An example will help to clarify the nature of this change. A relatively new outcome measure is that of health status, an individual's self-assessment of various aspects of quality of life and functional limitations (Ellwood, 1988). Health status measures have been used for specific diseases, but only recently have they been applied to broader populations. These measures may have a variety of uses, one of which could be to provide a snapshot of the health of a population. For example, a baseline health-status measurement could be applied to a targeted population, say health plan members between the ages of 30 and 50 who have no chronic diseases. Health promotion interventions could be developed for the group such as programs for weight reduction, cardiovascular conditioning, or smoking cessation. Health status could subsequently be remeasured so that a conclusion could be drawn about the impact of these programs on the health of the group.

Consider yet another example that relates to patient education using interactive technology. A patient is scheduled for an orthopedic procedure. The outcome of the procedure will be determined in part by whether the patient performs certain rehabilitation exercises after surgery. Under the old paradigm of care, whether the patient successfully performed the exercises was only of secondary interest (if at all) because compliance with rehabilitation was not viewed on par with such factors as infection rates, major complications, and other medical outcomes of the procedure. However, a different story emerges when a broader definition of outcomes is employed and health status measures are collected. Through these measures, a number of patients may be detected with a less-than-desired functional status after the procedure. The entire process of care is then evaluated to better understand the factors that contributed to these differences in outcome. After careful study, suppose it was determined that patients did not understand the importance of doing the rehabilitation exercises as recommended. An intervention like an interactive educational program could be used to explain the purpose of the exercises and to better instruct patients on how to perform them. The result presumably would be significantly better outcomes, a reduced need for further medical intervention, and, thus, lower costs and better quality. This simple example demonstrates how a new measurement combined with new technology for patient education could make significant improvements in the process of care.

WHERE ARE WE?

It is often said that there is nothing as valuable as a good framework. An organization that has a well-articulated plan and an effective method in place to implement the goals of the plan provides an excellent framework to begin the dialogue around the use of interactive technology for patient education and health promotion. As previously discussed, these plans allow many organizations to shift paradigms about what health care is about and therefore shift their thinking about the role of the patient, preventative care, and new technologies for health care services.

A major theme in this chapter is that building on existing activity within the organization can be an effective strategy. Many organizations are conducting what has been termed operational research. Contrasted to traditional clinical research, the research questions posed are directly linked to business goals. The research questions tend more toward effectiveness research rather than efficacy research. For example, a typical question might be, "Do shortened postpartum stays increase the morbidity of postpartum infection?" The answer to this question directly affects an operational issue: "How long should patients stay in the hospital after delivery?"

In a similar vein, it is not inconceivable to view computer technology for health promotion and the delivery of services as a form of medical intervention with patient satisfaction, compliance, and improved health status as the goals. The intervention could be framed as a research question requiring a careful study intended to answer a set of specific questions. This research approach will most likely also work in those systems with a more traditional research program.

Introducing interactive technology into a health care system will not be a simple matter. Cost containment activities are rampant in most organizations, and the amount of money for new programs is limited. Interactive technology still represents a major opportunity for research, improved health, better decision making, and cost savings for the health care system. Understanding the environment within the organization will help to develop better strategies for introducing these technologies into the health care equation.

REFERENCES

Barry, M. J., Fowler, F. J., Mulley, A. G., Jr., Henderson, J., Jr., & Wennberg, J. E. (1995). Patient reactions to a program designed to facilitate patient participation in treatment decisions for benign prostatic hyperplasia. *Medical Care, 33*, 771–782.

Ellwood, P. M. (1988). Outcomes management: A technology of patient experience. *New England Journal of Medicine, 318*, 549–56.

Hindle, T., & Lawrence, M. (1993). *Field guide to strategy: A glossary of essential tools and concepts for today's manager.* Boston, MA: Harvard Business School Press.

Kaplan, S. H., Greenfield, S., & Ware, J. E. Jr. (1989). Assessing the effects of physician–patient interactions on the outcomes of chronic disease. *Medical Care, 27*, S110–S127.

Luthans, F. (1995). *Organizational behavior* (7th ed.). New York: McGraw-Hill.

Relman, A. S. (1980). The new medical–industrial complex. *New England Journal of Medicine, 303*, 963–970

Rivo, M. L., Mays, H. L., Katzoff, J., & Kindig, D. A. (1995). Managed health care: Implications for the physician workforce and medical education. *Journal of the American Medical Association, 274,* 712–715.

Street, R. L., Jr., Voigt, B., Geyer, C., Manning, T., & Swanson, G. (1995). Increasing patient involvement in deciding treatment for early breast cancer. *Cancer, 76*, 2275–2285.

13

Reflections on Health Promotion and Interactive Technology: A Discussion With David Gustafson, Jack Wennberg, and Tony Gorry

This chapter presents a summary of a discussion that took place on April 12, 1996 with David Gustafson, PhD (Professor, Industrial Engineering and Preventive Medicine and Founder, Center for Health Systems Research and Analysis, University of Wisconsin), Jack Wennberg, MD, MPH (Director, Center for the Evaluative Clinical Sciences and Professor, Department of Community and Family Medicine, Dartmouth Medical School), Tony Gorry, PhD (Vice President for Information Technology, Rice University), and two of the editors (Richard Street and Tim Manning). The discussion focused on several issues related to the implementation and effective use of interactive technology for health promotion, patient education, shared decision making, and the delivery of health services.

INTERACTIVE TECHNOLOGY AND HEALTH OUTCOMES

Richard Street

Research investigating the effectiveness of interactive media for health promotion and patient education has produced mixed results. It is safe to say that interactive technology is at times superior to and at times no better than more traditional media (such as videotapes and brochures). How do you account for these findings? Poorly designed programs? Inappropriate applications? What criteria must be met in order to improve the effectiveness of these programs?

Dave Gustafson

First, differences in the outcomes that we're seeing may in fact be real. For example, the effectiveness of our Comprehensive Health Enhancement Support System (CHESS) programs appear to be influenced by differences in how people use the various services (e.g., the facilitation and discussion groups, expert mail, information libraries) as well as how the people are trained to use the system. The vast majority of users use CHESS a lot and think CHESS is great. We are beginning to learn more about differences between people whose quality of life seems to be enhanced by using the program and those people who seem to get less from CHESS. It's not the people who use CHESS the most; it's the people who use it the most intentionally. A person with pain may use the key word command to find information on pain, write a letter to the expert about pain, and so on. They come to CHESS with a purpose, and are not simply there to see what's going on in the discussions.

Second, I also think all of us are still learning how to effectively implement these interactive media. Some users need more instruction than others about how to use the program most effectively. It all comes down to meeting customer needs. The first use of the program needs to help users deal with issues that concern them at that time. If it satisfies these concerns, you have a committed user.

I guess another possible explanation is that we may be seeing a situation where there are some problems with the outcome measures themselves and the methodology for carrying out the evaluations of the programs. I think we have a long way to go in terms of coming up with adequate measures of effectiveness. In short, variability in outcomes in part relates to differences in patterns of use, in part to problems in implementation, and in part to problems of measurement.

Jack Wennberg

Many common conditions such as angina and benign prostatic hyperplasia can be treated in several ways. The choice of treatment really should depend on the preferences of the patient. Our purpose in designing the shared decision-making programs was to involve the patient in the actual decisions that are made with regard to a specific clinical problem, such as whether to have surgical treatment versus medical treatment for prostate disease. The kinds of obstacles we run into are not coming from the patients at all. In fact, our experience with randomized trials and cohort studies indicates that patients generally are enthusiastic about taking responsibility for decisions where their own values are (or should be) the important determinants in the choice of treatment. They feel more empowered by being brought into the decision process, and their satisfaction is greater when they are actively involved in decision making than when they are not. All of which is a very strong plus for the idea that, from the patient's perspective, this reform of the doctor–patient relationship is something that is much desired. The resistance that we have encountered has been the reluctance of providers to enter the shared decision-making contract and to give up the authority that traditionally has been vested in the provider to make the choices for the patient.

There are other troublesome issues, too, that pertain to the multimedia approach. Today, the issue of whether these programs should be done in an interactive media or in the traditional linear videotapes or brochures is an open question. In the future, however, as the interactive technology becomes more accessible and cheaper to use, its versatility is almost certainly going to assure a permanent role in this area. At this time, however, videotape represents logistically a much easier way of getting the shared decision making reform spread more widely. For one thing, videotapes can be distributed in many different ways. For example, when our shared decision-making programs have dealt with diagnostic screening tests, we have been able use videotapes to get them directly to the critical populations without having to bother the physician with remembering to prescribe the video.

A technological problem that we have encountered is the logistics of setting up self-standing multimedia platforms in the clinical environment. Another problem has been the intrinsic instability of these platforms at this time. I believe the technological solution to these problems is still a few years off. Nevertheless, I think that interactive multimedia will one day be in the home and in the clinic, and it will be part of the of the daily life.

Tony Gorry

I'm somewhat naive with respect to how this technology is actually used for health promotion and patient education. These are not areas that I work in directly. I would say, however, that the reaction that people have to media is much more complicated than those of us in the computing business would like to think. I'm presently involved in teaching about the role media plays in our perceptions of ourselves and the way in which we think about the world. It's obvious to me that not only are we inclined to gloss over differences in the way people react to media, but we're also somewhat insensitive to the ways in which media change what constitutes knowing about something. At some point, people's reactions to these patient education or shared decision-making technologies will be wrapped up in epistemological kinds of questions. I think the way in which people understand the world through these interactive computer programs is quite different from the way in which they might understand the world where they talk directly to a physician or someone else.

An example: Consider something like testing for acidity. In the past, people actually tasted the stuff. Now you can put them in a room with a computer screen, and on this computer screen is a representation or an indicator of acidity. It's different, but it's the same in the sense that many of us would say: "Of course it's the same." The computer screen is showing PH or whatever you want it to show. For the people involved, however, the experience and understanding may be very different. It may be so different that they themselves don't even understand it. This is a very simple example, but it illustrates the fact that although the information is the same, the people will experience it differently.

Now if you take something as complex as trying to understand what's wrong with my health or what I should do about my health, it's possible that the way in which the media itself impinges on the process may in fact change the way in which we understand the phenomenon. For some people, the shift in presentation from a physician to a multimedia program may be a step in the right direction. For other people, it may be confusing, disturbing, and may in fact not be understood. The presumption is that these programs will help people understand things better than they currently do. But it is not necessarily true or uniformly true.

At this point, I can't put a finger on detail. I just know that, during the history of major changes in communications technologies, one sees, at every juncture, shifts in the way people think about things. So it remains to be seen how this will play out with respect to whether people use these programs successfully, whether they really learn things, whether some people learn better than others, and whether decision making ultimately is improved. It is certainly a rich area to look into from a theoretical perspective.

Tim Manning

Let me give an example of what Tony may be talking about regarding epistemological issues. When asking questions about new technologies, we tend to form the questions in terms of the conventional processes and protocols we have for those processes. For instance, when you use interactive media to do patient education, you find that what you are doing changes so that you are not really doing patient education in the sense of presenting information that the patient should know. Rather you have connected patients with various kinds of resources and empowered them with an interactive interface so they can explore those resources. That's very different from educating patients by having them watch a linear program on information we want them to see.

We are currently developing a project that will enable patients to have access to physicians through e-mail or postings of various kinds. Patients also will have access to some portion of their medical records so that they can provide data, complete health status reports, keep a diary of their symptoms, and so forth. By allowing people access to health information, access to physicians, and the ability to cocreate the medical record, there may emerge a very different kind of relationship between the doctor and patient. Health education and health services may become more of a partnership and more patient-centered than in the conventional mode, where the patient visits the physician who makes observations about the patient's health, makes judgments about what the patient should do, and unilaterally makes entries into the medical record.

Jack Wennberg

I think it is also important that we not underestimate the importance of content. For example, in our research, we have found that global ratings of the media presentation by patients are independent of the effect of message content on the patient's preferences and understanding of the situation. In clinical trials, we have compared our videotapes to videotapes sponsored by pharmaceutical companies on the use of PSA (prostate specific antigen) testing. People seem to like the videos no matter what they say. However, people have very different behavioral responses, and their decisions are quite different, depending on the content of the video. To me, the key issue here is the problem of framing and the question of how knowledge or truth should be represented.

THE IMPORTANCE OF INTERACTIVITY

Richard Street

Developers of health promotion and patient education programs often presume that the interactive capabilities of computing technology potentially make it a superior method of delivery than are other noninteractive media. However, our experience indicates that interactivity sometimes is a positive feature of the media, and sometimes it presents problems. At issue here is how much interactivity is desirable. On the one hand, interactivity allows the user to explore the information in accordance with his or her individual needs. On the other hand, a highly interactive program may be confusing, frustrating, or provide an uninterested user the opportunity to skip the information. What are your thoughts on the concept on interactivity and how we should look at it in the development of these programs?

Jack Wennberg

We have used interactivity in a fairly low-key way. In fact, many people would not consider our programs to be very interactive. The interactivity comes in getting the right message to the right patient. In other words, we've used interactivity to access the data-base that applies to an individual patient's subgroup depending on information he or she has entered into the program (e.g., age, related illnesses). The program then presents scenarios particular to that patient's group in a seamless way as if he or she were seeing a fairly linear video. So we use the term interactivity very differently than other people might use it.

We have been more reluctant to develop programs that allow the individual lots of options of choosing and selecting and bouncing around in the program. Some earlier programs that tended to emphasize the interactivity were often not as successful as those that were more modest in that dimension. Patients were sometimes confused by the interactivity. That's my narrow experience. This, of course, has nothing to do with the wider question about the patient's ability to act interactively with information on the Internet or a computer network. That's a very different issue.

Tony Gorry

In principle, of course, interactivity is to be desired because it puts the user of the technology in a more active stance with respect to it. Interactivity presumably can facilitate understanding because of the ways in which people can pursue those aspects of the topic that they don't quite understand. I don't see any theoretical objection with respect to using interactive technology to get health care information to people. The obvious concerns, of course, are the seductive aspects of the technology and the extent to which different points of views, biases, or inclinations are embedded into the technology. I don't mean to sound ominous, but it's certainly true that if advertisers were interested in selling something to someone, then interactive and multimedia technology could allow them to shape the message and the presentation of the message to particular individuals based on their interaction with the technology.

So interactivity is a positive, but I think it also raises the stakes because of its seductive qualities. There's relatively little time to reflect on things in these interactive environments. It's not like taking pamphlets home where you can think about it and think about it again. It's quite possible that you can come through an interactive multimedia session having one opinion, but then later have second thoughts or perhaps even a different opinion. These are just general comments about the media and not so much about patient education or shared decision making.

Dave Gustafson

The devil is in the details, not in the basic principle. For instance, as we added more and more information to CHESS and gave people more and more options for navigating through the information, we found that some users became confused and tended to use the system less. So we made some significant changes. We now have a series of guided tours that are tailored to the most common concern that a user might have. For example, in our breast cancer program, we have an interface that allows a first-time user either to browse through articles of interest in the instant library or to take one of several guided tours tailored to particularly important and prevalent needs. We choose a subset of the program to view such as articles, questions and answers, personal stories, decision analyses, and so forth. In these tours we even suggest the order in which they see them. We found that many users need or want that guidance.

Second, some interactive programs run the risk of being cute with the buttons and icons that users must click on to work through the program. Often icons do not come with an explanation of where the person is going or what will happen if they click on it. We just assume that the word or the icon is self-explanatory. We find that people often don't look at those cute little icons on the menu bar and, if they do, they often don't understand them. We need to write out in English what is going to happen, what the options are at this particular point, and where they can go. The point is that you need to be real careful about how you design your interactivity or you will be causing more problems than you solve.

INSTITUTIONAL BARRIERS TO ADOPTION
AND IMPLEMENTATION

Richard Street

Health care institutions have been rather slow to invest in interactive technology for purposes of health promotion, patient education, and delivery of health services. There are probably a variety of reasons for this, including a limited number of programs, providers' attitudes about preventive care, institutional inertia, attitudes toward computing, to name a few. What do you see as the barriers to the institutional adoption of interactive technology for health promotion and patient education? How do we overcome these barriers?

Jack Wennberg

When trying to understand inherent resistances to change, a distinction needs to be made between resistance to the technology itself and resistance to ideas that are conveyed through the technology. First, there's an intrinsic resistance by providers to adopt the innovation—shared decision making in our case. With our programs, it's not so much the technology issue as it is the idea that the patient, rather than the doctor, should make medical decisions or at least participate in decision making. This is a transition that has been going on in our culture now for the last 20 years. Acceptance of patient involvement in care is slowly accelerating, and it's becoming easier to get this idea across.

There are, however, some very specific problems related to the technology such as the instability of the platform and the uncertainty about which technology an institution should invest in. We have a lot of people who really liked the Sony laser disc platform. However, they kept thinking that CD-ROMs were coming in the near future, and they didn't want to invest $8,000 in something that was going to be obsolete in 6 months.

Richard Street

Do you think the reason why health care providers, worksites, and patient groups are slow to invest in CD-ROMs and other stand-alone programs is because they look to the Internet as the medium through which these types of programs can be delivered?

Jack Wennberg

I think the whole situation is confused because there's so much uncertainty about where will the technology will go. Six months ago, we didn't even know much about the Internet, at least compared to what we now know. Even then the development of CD-ROM programs was slow to move. The CD-ROM technology that is available now is inferior to the laser disc platform that we had with Sony. We were reluctant to convert to CD-ROM although we thought we may need to because the laser disc platforms were no longer tenable. Now it seems that once again, we are waiting. This time we are waiting on the Internet to be sufficiently broad-banded so we can get the programs across it. Perhaps someday there will be more interactive television as well. This technology is becoming cheaper and more readily accessible at home and in the workplace. I think broad-scale success of our shared decicion-making programs will come when information is easy to access and cheap to transmit.

In thinking of ways or strategies to optimize use of the technology, it is important to understand that the primary customer here is the patient and not the health care institution. Our marketing interests now are trying to get the employers and representatives of patients to put pressures on the institutions to adopt this innovation. I think the patient's demands for information and involvement are increasing, and I'm fairly confident that in the future we will see less resistance by health care institutions.

Dave Gustafson

We currently have a project in a 5-county area surrounding Madison, WI, where we are trying to get the CHESS breast cancer program out to every woman over the age of 65. One of the major challenges we're facing (and one of the reasons for the project in the first place) is that there are many different physician and health care delivery organizations. Some are large multispecialty clinics; others are small, independent practices. The biggest barrier is not that physicians don't want CHESS. Virtually every surgeon and oncologist in that 5-county region seems to think that CHESS is a great idea, one they are happy to support. Yet, even the most supportive physicians have to change their behavior in order to optimize the use of a new technology.

We need to find a way of taking that responsibility off their back. Although health care providers may be enthusiastic and supportive of this technology and service, it will still be difficult for them to change their behavior. Let's keep provider behavior change to an absolute minimum. For example, identifying the patient in a timely fashion is our biggest problem. We may need to find a technological solution to this by integrating health support technology (like CHESS) with computerized medical records or other computer-based information systems that are already out there. For instance, is there a way to link into the clinical laboratory database so that, when a tumor is found to be cancerous, the people delivering the technology can find out without having to wait to hear from the doctor?

I also think we may be working with a wrong assumption. We tend to ask how are we going to get health care institutions to deliver this. I think that they have a major role, but we also need to offer patients the opportunity to get these systems through other routes. For instance, we are currently designing a project where community information centers are used to offer CHESS. In this project, underserved women in Madison are going to be introduced to CHESS at community centers. If they want CHESS in their home, they will go to a community center and not to the health care institution. The community center will train them to use CHESS, provide them a place to use it privately and, if they wish, arrange to have the technology placed in their home. We must increase the number of access points to systems like CHESS. We cannot rely on health care providers as the sole point of access.

Tony Gorry

Let me make a couple of comments on these issues. First, inferior technology does not necessarily lose out. All around us, we see technology that is successful but not necessarily a step forward. A primary example would be the technology underlying the World Wide Web, which has been phenomenally successful. The technology is quite old and in some ways a step backward. It is the way in which the technology gets absorbed and used that matters. It may well be that the CD technology will become common. Certainly we understand people's willingness to rent scratchy old movies at the video store. That technology is decidedly inferior to much of the television technology, yet it is very successful.

The other thing, though, is that health care institutions generally are not very innovative. There is an inherent conservatism in the organization that is generally justified on the grounds that we are taking care of patients and lives are at stake. This way of thinking makes it easy for administrators and doctors to justify doing business as usual. I think cost pressures also make it easy not to invest in innovation, at least innovation that doesn't directly relate to medical treatment or diagnosis. There are some exceptions, of course. Nevertheless, it will take some major wrenching of medical centers, hospitals, and the like to get them to do much of anything that is innovative. They will talk about innovation, but many will not do much.

I think that as the demand for this technology increases, hospitals and providers will have to do something. Whether that something will involve patient education or health promotion also will depend on who is putting on the pressure. Although there are technological reasons why this technology is not widely used, most are organizational, attitudinal, and cultural.

Of course, one way to overcome this resistance is with successful projects. People often are influenced by someone's ability to do something. However, there still needs to be a major shift in the attitudes of hospitals and doctors. Historically, they have had no strategic interest in people being healthy or highly informed. This may change with capitation and managed care because now there is a motivation for providing services that keep people healthy. I still think it's going to take awhile for all of this to sink in and for people to change.

On the practical side, the first order of business is to get people in positions of leadership to understand that there are in fact benefits to using interactive technology for health promotion and for delivering health services. Those benefits will have to come out of the work that Jack and others are doing. I'm not familiar with all the work, but I've seen examples and interesting demonstrations of the power of this technology. Nevertheless, it has to be put into an organizational context to make it work.

Jack Wennberg

I agree that capitation is the key to getting the financial incentives right. But it's more than just keeping people well. It's making certain that the institution is not accused of withholding the care that people want. Often, that care comes in the form of elective surgery. Because of the strange variations in surgical rates around the country, there is a concern that we have underservice in some places and overservice in other places. However, the patterns of variation today reflect the peculiar distribution of suppliers and how they, and not the patient, are deciding what elective surgical procedures should be performed. Once the patient is empowered to choose, surgical rates would be determined by patient preference, and now we have only an inkling of what the demand curve would look like. In some of the HMOs where we've implemented our programs, surgical procedure rates dropped. This suggests that patients are more risk averse than surgeons. As information becomes available and as capitation becomes the dominant mode of financing, health institutions will need to persuade the public that they, the health care provider, are not inducing demand as they had in the past. In a capitation environment, all the incentives go on the cheap side. Now the institutional attitude may be "Don't do anything" when in the past it was "Do a lot." But that kind of new thinking implies withholding of care or at least the threat of withholding care. The solution is to adopt shared decision-making to make it possible for patients to set the level of demand. I think the struggle to assure the public that capitation does not lead to withholding of valued care will work in the favor of overcoming the barriers to adoption of programs designed to involve patients in medical decision making and staying healthy.

THE FUTURE OF INTERACTIVE TECHNOLOGY
FOR HEALTH PROMOTION

Richard Street

What do each of you see as the future of interactive technology for health promotion and patient education purposes? If you have ideas for the best case and worst case scenarios, I'm sure the readers would be interested.

Tony Gorry

I should start by saying that we have the technologic potential to do more than what organizations are doing currently. I firmly believe that 5 to 10 years down the road, we will have enormous technology at our disposal. We will have high band width, very fast computers, and DSP technology that will enable us to do mixed video and other things together on the work station. We will have the equipment for interactive television, whether it's an actual television or a PC. All these things will be possible, but the real question is whether the people in organizations will envision the ways to use that technology effectively. My guess is the answer will be mixed. Some organizations will; many will not. Those that will not will be more timid and or absorbed in other kinds of matters. They will wait to see how things work out with the more innovative groups.

I think an interesting thing that's happening in the absence of medical institutions taking a leadership role is that there is a cottage industry developing in medical information. If you casually browse the Internet, there is a sea of uncritiqued, unedited information for virtually any medical problem. If you actually pick a subject and present it to one of the World Wide Web search engines, it will come back with a lot of material. Much of it is probably junk. That is what people will get unless medical institutions get more aggressive with how they organize information, how they get it to people, and how they involve people in being good consumers of information and good judges of the information's validity. If medical institutions and health agencies do not get involved, it will become a kind of carnival of medical information that may ultimately undermine the credibility of the established institutions. It may have the same effect as television has had on family life. There will be so many different lifestyles and different opinions that it will be hard for the traditional medical information resources to hold sway.

If institutions want to encourage lifelong kinds of commitments with their patients, then they will have to reach out to those patients because it is no longer the case that patients only get their information from medical institutions. Good information that is well presented, well organized, and thoughtfully integrated in the patient's life seems to be strategically one of the most important things a medical institution can do. So it strikes me that, if leaders of medical institutions want the institution to prosper in the future, they should start dealing with patients as real customers, not as units to be moved through some logistical system at the lowest possible cost.

Jack Wennberg

I think that is a very interesting perspective. I have been very concerned about the breakdown in the authority structures in medicine. The authority structure as embodied in scientific review gave us some guarantee that information had gone through some process of validation before it was broadly distributed. Much of my career has been involved with critiquing the scientific validity of what passes as normal science in medicine, and I must say that some of the scientific claims of the orthodoxy often cannot be substantiated. However, a process has been in place to work these inconsistencies and false claims through to their resolution. The opportunity for false claims gets magnified once unscreened information is broadly available. Much of the orthodoxy has focused on narrow technical issues of efficacy, leaving aside the issue of whether end results of a biomedial intervention are what patients actually want.

I gather from Tony's remarks that we may begin to see real speciation in schools of thinking that will become new threats to orthodoxy. I'm not sure that's all bad, incidentally, because the conventional system has been driven by a technological imperative regarding treatment that seems to be more and more out of tune with what patients actually want. Consider, for example, the continuing debate in this country about the mode of death as well as the role of patient preferences in the decision to undergo elective surgery. Perhaps we are in for quite a ride here. It is also clear that medical institutions can no longer maintain the tradition that has the provider making the sole judgment about the value of medical care.

Tony Gorry

Leaders of American medical institutions ought to get out on the Internet and see what's passing for medical information. Much of it is implicitly and sometimes explicitly critical of existing medical orthodoxy. I think it will be in the strategic interests of institutions to provide good information. I think good information will be the element that binds people to institutions and maintains their sense of trust.

Richard Street

Let me pursue that line of thinking a little further. We have this struggle to some extent in the review process for scientific work. Even in a peer review process, we often disagree on what's quality and what's not. We have conflicting findings such as whether coffee is bad for you or not. Add to this the burgeoning growth of health information on the Internet, most of which is not reviewed. It seems to me, Jack, that you are saying that this may be a problem, but again it might also provide an opportunity for different perspectives to come out on those issues.

Jack Wennberg

It's the value side too. The traditional biomedical establishment has created a lot of problems for patients. Now I haven't had the chance to look through the Internet as much as Tony has, but I've looked under *prostate cancer treatment* because I am interested in that topic from a scientific perspective. You can see all sorts of different representations of what reality is out there on the Internet. Patients are desparately seeking solutions to their predicaments. Many seem quite angry that there are no clear answers—but many want to take responsibility for the decision making, even when the science is weak. It certainly is in this case. Interactive technology may encourage a tremendous dissolving influence on the paradigm of medical care we've been living in for the last 25 to 30 years.

Dave Gustafson

Let me share some of own experiences with this issue. In the Alzheimers area, we have been monitoring the Internet discussion groups periodically over the last year. During that time, there appears to be a dramatic transition in the way in which people are talking about various issues in these groups. A couple of things appear to be happening. First, more and more of these groups have become guinea pigs to be studied by researchers of one type or another. Also, these groups are attracting an audience for people who feel very strongly about certain issues. As a result, the participants are becoming much more closed, much less willing to share, and much more suspicious of each other.

I think we are only beginning to get a hint of where the Internet is going to take us. Taking Tuckman's model of forming, storming, norming, and performing, use of the Internet is in the forming period where we are really excited about the possibilities. We're now starting to see people get frustrated and depressed (storming) because of the problems they are encountering. I expect we will slowly work our way out of this to a more realistic assessment of what this technology could be and can do (norming). But right now we are not sure how this is going to play out. In the worst case, we're going to have a sophisticated system that is controlled by charlatans, quacks, researchers, and snoops. Unless there is a way to control access to key elements, the Internet will become viewed with suspicion, and it will have very limited use as a vehicle for open communication. It may become a series of Intranets that will have very limited and very controlled access. If that happens, there will be a lot of disenfranchised people. These will be the people who don't have easy access to communication networks either because of finances, a lack of knowledge, or whatever. Right now all sorts of materials are free on the Internet. I doubt that the *Wall Street Journal* will be free on the Internet 5 years from now. I think our free access to many of these services will become more and more limited. As time goes on, there will be more and more people who don't know what's out there; don't know how to access it; don't have the resources to access it, and even if they had the resources, they don't trust it anyway.

On the positive side, I think that we are going to see a lot of improved technological capabilities. Right now, it's hard for the Internet to support some of the expert systems on CHESS. However, with Java and other resources coming downstream, the Internet will be able to deal with expert systems and other more sophisticated applications of computers. I think the technological limitations are only a temporary problem. As the technology becomes more sophisticated, the potential becomes much more real for computers to provide a very effective means for supporting people in need.

However, if we are not careful, the gap between the haves and the have-nots, between the information-rich and the information-poor is going to grow dramatically. It will be yet another example of what we're seeing in the rest of our society, the elimination of the middle class and the continued marginalization of the poor. We must find a way of getting our systems into the homes of the underserved.

AUTHOR INDEX

SUBJECT INDEX

Milton Keynes UK
Ingram Content Group UK Ltd.
UKHW031135141024
449569UK00006B/172